Synergy:
Science Reasons
With
Atheists

Steve Kellmeyer

To obtain additional copies of this book, contact:
www.bridegroompress.com

Copyright
March 2010
Bridegroom Press

Printed in the U.S.A.

ISBN: 978-1-60104-034-3

Contents

St. Augustine of Hippo

Usually, even a non-Christian knows something about the earth, the heavens, and the other elements of this world, about the motion and orbit of the stars and even their size and relative positions, about the predictable eclipses of the sun and moon, the cycles of the years and the seasons, about the kinds of animals, shrubs, stones, and so forth, and this knowledge he holds to as being certain from reason and experience. Now, it is a disgraceful and dangerous thing for an infidel to hear a Christian, presumably giving the meaning of Holy Scripture, talking nonsense on these topics; and we should take all means to prevent such an embarrassing situation, in which people show up vast ignorance in a Christian and laugh it to scorn.

The shame is not so much that an ignorant individual is derided, but that people outside the household of faith think our sacred writers held such opinions, and, to the great loss of those for whose salvation we toil, the writers of our Scripture are criticized and rejected as unlearned men. If they find a Christian mistaken in a field which they themselves know well and hear him maintaining his foolish opinions about our books, how are they going to believe those books in matters concerning the resurrection of the dead, the hope of eternal life, and the kingdom of heaven, when they think their pages are full of falsehoods and on facts which they themselves have learnt from experience and the light of reason?

Reckless and incompetent expounders of Holy Scripture bring untold trouble and sorrow on their wiser brethren when they are caught in one of their mischievous false opinions and are taken to task by those who are not bound by the authority of our sacred books. For then, to defend their utterly foolish and obviously untrue statements, they will try to call upon Holy Scripture for proof and even recite from memory many passages which they think support their position, although they understand neither what they say nor the things about which they make assertion.

~ *The Literal Meaning of Genesis.* vol. 1, *Ancient Christian Writers.*, vol. 41, Translated by John Hammond Taylor, S.J. , 1982

For a long time I wanted to become a theologian ... now, however, behold how through my efforts God is being celebrated in astronomy.

~Johannes Kepler

Mathematician, astronomer and Lutheran.
Improved the telescope and developed the laws of planetary motion.

Christianity - that is, the religion of the Bible - is the only scheme or form of belief which disavows any possessions on such a tenure [that is, it is the only religion that denies taboos]. Here alone all is free. You may fly to the ends of the world and find no God but the Author of Salvation. You may search the Scriptures and not find a text to stop you in your explorations.

James Clerk Maxwell

Theoretical physicist, mathematician and devout Presbyterian. He contributed to the fields of optics, color vision, gases and elasticity but is most well-known for his work in uniting electromagnetic theory with light theory. His work is the basis for all modern physics. He was also a devout Christian deeply familiar with the Scriptures.

Introduction

1. What is the difference between experimental science and theology?

Experimental science investigates, discusses and measures the relationships between inanimate objects. The science of theology investigates, discusses and explains the relationships between persons.

Experimental sciences do not explain the relationships between objects. At best, they describe these relationships using measurements and words like: "mass," "velocity," "energy," etc.

Theology both describes relationships and explains those relationships. However, because it deals with persons, it does not measure at all. Instead, it uses personal terms such as "love," "hate," "faith," "intimacy," "hope," etc.

According to theology, both the inanimate universe and the persons living in the universe tell us about God. Both kinds of knowledge are necessary for our understanding of God, both Who He Is and Who He Is not.

2. Is everything that happens God's Will?

Yes, in the broad sense, everything that happens either reflects God's permissive will or His ordaining will. Distinguishing between the two is critical to understanding how the world works.

God created all things good, He is never the author of evil or sin, yet there is sin and evil in the world. So when we say that everything happens according to God's will, we have to distinguish two different ways in which His will operates. We must see the difference between His "ordaining will" and His "permissive will."

His ordaining will is what He ordains for us, that is, what He intends for us. God only intends the good for us. He calls us to the good, He gives us everything we need to embrace it. He wants us to be happy.

But, we have the freedom to reject the good. That is, God permits us to reject Him if we really want to. God is free. We are in the image and likeness of God, so He allows us to act freely. He does not constrain us. This is the basis for discussing His permissive will.

When we choose good, the true or the beautiful, we choose freedom. We choose God. When we choose against the good, the true or the beautiful, we choose slavery. We choose against God.

> The gift of mental power comes from God, Divine Being, and if we concentrate our minds on that truth, we become in tune with this great power.
>
> *~Nikola Tesla*
>
> *Serbian Orthodox mechanical and electrical engineer.*
> *He invented the AC system of power generation and transmission.*
> *The Standard International unit of flux density is named for him.*

How does this work in practice? God holds all that exists in existence from moment to moment. The subatomic particles, the atoms, the molecules of the chair I sit on, the chair itself, is held in existence by God. I can sit in it, or I can pick it up and beat my neighbor to death with it. If I choose to do the latter, God continues to hold the chair in existence while I beat my neighbor to death with it. He doesn't interfere with my decision to misuse the chair. He doesn't cause it to pop out of existence in order to prevent my misuse of it.

This radical ability to do exactly as we please with the things God has given us in the created world is God's permissive will. He gives me things - wood, nails, glue - which I can use to create a chair so I can give myself or another ease and comfort with it. He intends me to use it as He would use it – to help others. But, if I choose instead to use those materials to distort or bring evil into someone's life, He will permit me to do it.

He doesn't desire me to harm others, but He empowered me to be His co-worker when He brought me into existence at the moment of my conception. He expects me to live up to my responsibilities as a divine co-worker. If I do, I will find myself more filled with joy than I could possibly imagine. He entices me with the good. But, He allows me to experience how refusing the good can harm both me and the world. If I don't live up to those responsibilities, the world is so constructed that I will find myself very unhappy.

The world is also so constructed that no matter how much evil I attempt to wreak within it, no matter how much I refuse the good, God can use even my refusals as a way to entice everyone towards the good. Precisely because He knows how to set every single thing to an infinitely superb pitch of rightness, He can afford to allow us our puny mistakes. Each of us can even be allowed to choose an eternity of refusal, if we really wish. He doesn't desire us to do that, but He permits it out of respect for us, respect for what He has made. So, everything is God's will, and everything is also my choice.

3. Does God cause natural disasters?

No, God does not cause natural disasters. We do. Now, this doesn't mean I personally caused the hurricane which destroyed my house. It means mankind has, through our own free choice, placed ourselves in a difficult position.

Consider an avalanche. If an avalanche happens in an uninhabited valley, it is not a disaster; it is merely the movement of some physical objects from one location to another. Indeed, seeing an avalanche, a lightning strike, a tornado, a tsunami, a hurricane or any other display of nature's power is exciting and awe-inspiring. None of these things are considered disasters in and of themselves. They are only called "disasters" when human persons, or the various objects to which we are strongly attached, happen to be harmed during the course of the event.

Before the fall of man, mankind was both immortal and impervious to suffering. Further, he did not have an unusual attachment to created things. Consequently, in an unfallen world, if unfallen man or any of the things he used were caught in any of these events, no disaster would result.

He wouldn't suffer any loss. He couldn't be injured by them; he couldn't be killed by them. If his things were destroyed, he wouldn't consider it a problem because he knew God would provide more. God would either do this directly or, using the innate God-given power the man himself had, the man could replace the lost things himself. No loss, thus no disaster. In short, it is not the natural event that has changed, but man's perception of what it does, his reception of how it affects his life.

> Of all created comforts, God is the lender; you are the borrower, not the owner.
>
> **~Ernest Rutherford**
>
> *Chemist and Physicist.*
> *He discovered "alpha," "beta" and "gamma" radiation, was the first man to transmute one element into another (nitrogen into oxygen) and predicted the neutron. Element 104, rutherfordiam, is named in his honor. He is called the Father of Nuclear Physics.*

We are only adversely affected by natural events because our father and mother, Adam and Eve, chose to reject the power to remain unaffected. Our great-great grandparents didn't want God's grace, His protection, so out of politeness to us, He doesn't extend as much protection to us as we would otherwise get. We are no longer immortal. We are no longer impervious to harm.

Now, God is very generous, so He generally protects us from many things anyway. Still, we are able to experience pain and loss to a degree we are not really equipped to handle. How can this problem be solved?

God solves it by allowing pain and loss to become a means by which we can be made whole. It would be better if this were not necessary, but we weakened ourselves through sin. We have trapped ourselves into this situation.

As was earlier pointed out, God gave us radical freedom and He respects how we choose to use it. But, through the suffering and death of Christ, God changed the nature of suffering and loss. Suffering has now taken on its own salvific power and meaning. Suffering, when joined to Christ's own suffering and death, can now heal the world.

Pain is still a natural evil. It is not good. As a result of the fall, we gained the ability to suffer in extraordinary ways. But, as a result of the Cross, we gained the extraordinary ability to turn this suffering into work that helps to heal what's wrong with the world. The natural evils which are suffering and death can be used as a conduit through which God pours in grace and repairs the damage caused to the world. Like a martial arts expert, God took what looked like loss and uses its own power against it to steal victory.

So, God doesn't cause natural disasters. Instead, we are subject to them - that is, they are capable of being disasters - because of who we are. But, if we permit Him to do it, God can use those disasters to change us. When we are fully changed, such events will again no longer be disasters. Death will be swallowed up in victory.

4. Can science prove the existence of God?

Experimental science is a tool; it is a way of thinking about inanimate aspects of reality. Experimental science only deals with physical objects. It cannot deal with anything which is not a physical object. God doesn't have a body, He is not a physical object. Thus, experimental science cannot prove the existence of God.

The science of theology, on the other hand, most certainly can prove the existence of God. Before we study any of the proofs theology may supply, however, we have to recognize that theological proofs are going to look quite different than the proof experimental science requires, if only because persons are not objects, nor or objects persons.

Remember, God is not the only thing experimental science cannot adequately handle. Justice, mercy, evil, love, hate, hope, beauty, faith, angels, demons: experimental science can't prove any of these exist either. If it cannot prove the existence of these things, does that mean none of them exist? Well, no. But it doesn't mean they *do* exist either.

Well, what does this failure mean? It means this particular kind of science is not capable of answering this particular kind of question one way or the other. It isn't an appropriate tool for the problem. We have to turn to a different kind of science to get answers to these questions.

Since theology is, indeed, a science, we can say with certainty that science can prove the existence of God. But we must remember that theology is a formal science, not an experimental science.

5. What is a formal science and how does it differ from an experimental science?

A formal science studies essential principles. These essential principles are not furnished by the natural world, rather, they come from the very essence of a thing. A formal science has its own language, symbols and rules. A formal science starts with basic premises or statements and reasons from those premises. As long as the premises are true and the reasoning valid, the results will be true. Math, grammar, logic, law, theology: these are examples of formal sciences.

Experimental science, on the other hand, studies the things in the natural world. An experimental science starts with an idea of how some known object within the universe might work and then tries to strictly measure some portion the universe to discover if the original idea was correct. Chemistry and physics are experimental sciences.

Experimental sciences (like physics) require the support of formal sciences (like mathematics) in order to correctly understand the universe. Formal sciences do not *require* the support of experimental sciences to work. However, formal science is usually very much helped by the work of experimental or applied science.

> Religion and science demand for their foundation faith in God. For the former (religion), God stands foremost; for the latter (science), at the end of all thought. For religion He represents a basis; for science, a crowning solution towards a world view.
>
> ~ *Max Planck*
> *German Physicist and Father of Modern Quantum Theory.*
> *Planck was a churchwarden from 1920 until his death.*

Applied science uses human knowledge, whether that knowledge is gained through formal science or through experiment, to solve specific problems. Optics, electronics, and applied mathematics are examples of applied science. If theology is a formal science, then religion, the way we live out theological truths, can be considered an applied science.

So, while neither mathematicians nor theologians ever use test tubes, Bunsen burners or particle accelerators, they do use formal systems of logic to reach their conclusions. They are doing science.

Section 1: Cosmos

1. How old is the Universe?

The best estimates of the age of the universe come to us through the physical sciences. These estimates have varied as the precision of our tools have varied, and as our interpretation of the measurements those tools make have varied. Currently, the best estimates place the age of the universe somewhere between 13.5 billion and 20 billion years old, probably closer to the lower figure than the higher.

2. Does the age of the universe according to the physical sciences conflict with what the Bible says?

No. The Bible does not provide a numerical age for the universe, nor the age of the earth, the moon, or any other inanimate object, for that matter. The Bible isn't really meant to tell us any of those things. That's why Christians have always recognized that the natural sciences can be, in regards to certain subjects, more informative than Scripture. As St. Cardinal Bellarmine, a doctor of the Church, wrote to Galileo:

> I say that if there were a true demonstration that the sun was in the center of the universe and the earth in the third sphere, and that the sun did not travel around the earth but the earth circled the sun, then it would be necessary to proceed with great caution in explaining the passages of Scripture which

seemed contrary, and we would rather have to say that we did not understand them than to say that something was false which has been demonstrated.

Galileo phrased it more succinctly, "The Bible tells us how to go to heaven, not how the heavens go."

3. How did the universe come into existence?

As far as physical scientists can determine, matter did not always exist. The idea that matter always existed, that it had no beginning and no end, is called the Steady State Theory. Although this theory was very popular when it was developed by astrophysicist Fred Hoyle in 1948, most physicists now agree that the physical evidence points instead to an older theory, a theory which Hoyle had derisively dubbed the "Big Bang." Although he was making fun of this older theory when he used the phrase, the name stuck.

In 1948, Hoyle, an atheist at the time, had no quarrel with the physical evidence for the Big Bang theory, but the idea had been developed by a physicist who was also a Catholic priest. Consequently, many people felt the theory implied a Creator. Hoyle developed the Steady State theory in order to avoid the philosophical implication of a Creator. As an atheist, he found the Big Bang theory repugnant. So, the development of the Steady State theory was not driven by evidence so much as it was driven by Hoyle's specific (atheistic) philosophical outlook.

In this respect, Dr. Hoyle's rebellion against the Big Bang theory and the theistic philosophy associated with it was not tremendously different from today's evolution debate. Many individuals today have no quarrel with the evidence for biological evolution, but they dislike the atheistic philosophy which has become attached to it. Consequently, they attempt to develop alternate theories which seem better able to support a theistic perspective.

> The best data we have are exactly what I would have predicted had I nothing to go on but the five books of Moses, the Psalms, the Bible as a whole, in that the universe appears to have order and purpose.
>
> **~ *Arno Penzias***
>
> *Physicist and Nobel Laureate.*
> *He and Robert Woodrow Wilson discovered the background*
> *radiation which confirmed the Big Bang theory.*

It is interesting to note that Dr. Fred Hoyle did not remain an atheist. As he was working out the theory for how the various chemical elements came into existence, he noticed something peculiar in a particular carbon-producing reaction. On the basis of that very peculiar reaction, he made several predictions concerning the attributes of carbon. His predictions turned out to be correct in every respect. This shook him to the core, if only because this particular reaction was, from a statistical point of view, exceedingly unlikely. Hoyle later wrote:

> Would you not say to yourself, "Some super-calculating intellect must have designed the properties of the carbon atom, otherwise the chance of my finding such an atom

through the blind forces of nature would be utterly minuscule." Of course you would . . . A common sense interpretation of the facts suggests that a superintellect has monkeyed with physics, as well as with chemistry and biology, and that there are no blind forces worth speaking about in nature. The numbers one calculates from the facts seem to me so overwhelming as to put this conclusion almost beyond question. (Fred Hoyle, "The Universe: Past and Present Reflections." *Engineering and Science,* November, 1981. p 8-12)

As a result of his own scientific observations, calculations, predictions and their confirmations, Dr. Fred Hoyle became a Christian.

While Hoyle's predictions concerning the creation of carbon were astute, his Steady State theory did not fare so well. A host of discoveries, including that of the cosmic microwave background in the 1960s, the 1980's discovery of young galaxies and the 2001 discovery that the cosmic background noise is distributed unevenly all have confirmed the original Big Bang theory. This is as close as we are likely to get to understanding the beginning of the universe.

4. What is the Big Bang theory?

Developed by the Belgian astrophysicist and Catholic priest, Georges Lemaître, the Big Bang theory asserts that a massive explosion brought every created thing into existence at some point billions of years ago. Physicists do not know what caused this event, nor can they tell us very much about the massive expansion. Due to the nature of quantum physics, it is impossible to say what the universe looked like or what physical laws held true prior to about the first 10^{-43} portions of a second prior to the explosion. Due to the limitations of the physical universe we live in, we are only capable of observing the results that followed after that first short interval of time. Thus, the Big Bang theory really only starts after creation has already sprung into existence. In this respect, the Big Bang theory is not remarkably different from the theory of evolution.

> There is a kind of religion in science, it is the religion of a person who believes there is order and harmony in the universe, and every effect must have its cause, there is no first cause... This religious faith of the scientist is violated by the discovery that the world had a beginning under conditions in which the known laws of physics are not valid, and as a product of forces or circumstances we cannot discover. When that happens, the scientist has lost control...
>
> ### Robert Jastrow
>
> *Astronomer, physicist, cosmologist and agnostic.*
> *He was the founder of NASA's Goddard Institute and is now director of the Mount Wilson Institute and its observatory.*

The Big Bang theory tells us how the universe expanded and changed over the course of time, but it does not tell us how the universe came into existence or what, if anything, existed before it. Similarly, the theory of evolution attempts to describe how life changes over time, but it cannot and does not describe how life came into existence from non-life.

5. Hasn't physical science demonstrated that the Creation account in Genesis is wrong - that there couldn't be light prior to the creation of the sun?

Actually, if the Big Bang theory is correct, light would have to appear well before the sun did:

> Before the first galaxies, before the first stars, there was light – the brilliant glow of radiation created during the Big Bang, 13.7 billion years ago. The remnant of that ancient light, now cooled by the expansion of the universe to a frigid 2.7 Kelvin, is known as the cosmic microwave background. ("Planck by Planck," *Science News*, April 11, 2009, Vol. 175, #8).

Scripture is not a science textbook; it is not meant to explain astrophysics to us. Indeed, if we were to accept Scripture as an absolute, physically accurate description of the universe, we would be forced to believe the world was flat, with a dome above it. Waters would be above the dome, and rain would issue forth only because God opened the doors in the dome of heaven to allow the water to pour down. The moon would be in a higher dome, the stars of heaven would be in yet a higher dome. Thus, the whole world would look something like a tent pitched on a flat surface with the rest of the universe suspended above the sky (the roof of the tent).

This is obviously an incorrect physical model for the universe. Scripture is clearly not trying to accurately describe the physical relationships between various heavenly and terrestrial bodies. It is, however, interesting that the current 21st century standard model for how the universe came into existence does not disagree with the idea that light came into existence before anything else, an idea found in both the first phrases of Genesis and the Gospel of John. It is also interesting that on the largest scales, space is geometrically flat. However, from a theological perspective, these coincidences have no real bearing on Scripture.

6. How can God know our future when quantum mechanics shows no one can even fully know the present?

To understand this, we must first understand what quantum mechanics tells us. Fundamentally, the universe is made up of fields. The various subatomic particles that make up atoms, molecules, etc., are just manifestations of those fields. Particles are really just very high energy fields. Because everything is constantly in motion, and everything is really, at its lowest level, just a field, the physical qualities of the smallest field-objects, like position and momentum, cannot both be precisely measured at the same time. The more precise we become with one measurement, the more difficult it is to be precise with the other.

We have to realize that all of these things - the fields themselves, their position, momentum and measurement - exist inside of space and time. According to the best physical theories, both space and time are created things; both came into existence with the Big Bang.

We are inside and part of the universe, God is outside and not part of the universe. Just because we who are part of what is inside the box cannot easily manipulate things like fields, particles, space and time doesn't mean God Who Is outside the box cannot easily hold them in existence. There is an immense difference between manipulating something from inside and holding it in existence from outside.

For instance, when I read a book, the characters in the book cannot grasp hold of the words I am reading, the words that describe them. They cannot grasp the pages. But I, who wrote the book, or even I who merely read the book, I can jump to any portion of the book, I can bring the book into existence and hold it in existence.

God's relationship to this universe we inhabit is somewhat similar. He brought it into existence, He holds every aspect of it in existence, even time itself. From moment to moment, He can see every part of it simultaneously, past, present and future. Saying God can't do something because we can't do it is not rational.

7. **If God knows what will happen, why doesn't He stop the bad things from happening? Why is there evil in the universe? How can we be free? Doesn't His foreknowledge make everything in the universe, including us, just a machine?**

A parent who has a three-year old can answer these questions more easily. Very small children often-times get strange ideas stuck in their heads. They want to play with something that is definitely breakable, or play with it in a way that is definitely dangerous. If they do what they want to do, they will end in tears.

A parent who knows his or her child well can often "see" what the child is thinking. The father can warn the child not to do what the child is contemplating. The mother can explain why it's a bad idea. But the parents know what that child will do if given the chance. And, sometimes, parents must give the child that chance.

Knowing God without knowing our own wretchedness engenders pride. Knowing our own wretchedness without knowing God engenders despair.

~Blaise Pascal

Mathematician, physicist and Catholic philosopher. Invented the science of statistics, and contributed to understanding pressure and the vacuum. Late in life, he had a vision of God as divine fire. He was an ardent defender of the Real Presence of Christ in the Eucharist. The Standard International unit of pressure is named in his honor, as is a programming language, Pascal's law of hydrostatics, Pascal's Triangle and Pascal's Wager.

If the child is to learn how to distinguish a good decision from a bad one, he must be allowed to make the decision. So, the parent can predict what will happen no matter which course the child chooses. If the child chooses to obey, the parents know things will go well for everyone. If the child chooses to disobey, the parents know things will go badly all around. They know the future, but they don't choose to control it. They permit the child to make the choice.

God is in a similar position with us. He knows human nature much better than we do – He made us. He has told us how we can do well and how we can avoid suffering. He has also told us what things we should avoid, lest we suffer. He knows what will happen, no matter which course we choose. But He lets us choose.

God grants us a remarkably radical freedom. He made the universe as our sandbox – we can do what we want in it. Anything at all. We choose our own consequences.

Now, sometimes things happen to us that we do not choose – we get hit by a bus and end up in a hospital, or encounter cranky people who yell at us for no reason. But, even when faced with these uncontrollable situations, we still control how we will react to them. We choose how we will respond. We choose how to live. God freely established a wonderful universe for us. We must freely choose to live well in it.

8. Why did Christians believe in a flat earth?

They didn't. In fact, no one educated in Europe since the 2nd century BC has accepted the idea that the earth is flat. The idea that anyone in medieval Europe believed in a flat earth is a falsehood propagated by Andrew D. White (1832-1918). His two-volume work, *A History of the Warfare of Science with Theology in Christendom,* published in 1896, had the appearance of being scholarly because he included numerous footnotes. In fact, it was all a charade. He was an atheist with an axe to grind. He didn't write history, rather, he ignored all historical evidence contrary to his thesis.

But that left him with precious little evidence to support his claims. So, White didn't just ignore contrary evidence; he was reduced to lying, forced to crib from works of fiction when he felt the situation called for it. In this case, he stole a minor plot device from Washington Irving's 1828 fiction novel *The Life and Voyages of Christopher Columbus* and incorporated it into his "history." As a result of White's theft, many people still actually think Christopher Columbus and his sailors believed in a flat earth. They didn't. Columbus thought the sphere much smaller than it was, and so mis-judged the time it would take to make the voyage.

> We regard it as a certainty that the earth, enclosed between poles, is bounded by a spherical surface.
>
> **Nicolas Copernicus**
> *Astronomer, mathematician, canon lawyer and Catholic priest.*
> *He developed the heliocentric theory and assisted with the*
> *Gregorian reform of the calendar.*

Ironically, due to the enormous popularity of White's fiction a century ago, most modern Americans now believe the lie that medieval Catholics used to believe a lie.

9. Are there additional space-time dimensions?

No one knows. The idea that there are an infinite number of space-time dimensions, each its own universe with the possibility of an infinite number of life forms, is essentially impossible to prove by any known scientific method. A statement made either way is a faith-based statement.

Whether it is true or false has no effect upon the Catholic Faith. The Church exists in order to heal the broken relationship between God and man. The existence of other universes would not affect this primary task.

10. Can miracles really happen?

Oddly enough, this question is somewhat related to the question "Are there multiple universes?" Experimental science is only capable of studying the laws of nature as they are. We live in a fallen world. As a result of the fall, the universe we inhabit lacks the grace it is supposed to have. Grace is God's presence, it is God's power. So, our universe is not as powerful, which is to say, not as grace-filled, as it should be.

It is quite likely that an unfallen world would contain within it options and possibilities that are not available in a fallen world. Thus, we can look at a miracle as a breaking-through into this world of the rules that are available in an unfallen world. That is, one could argue that, through a miracle, we get a glimpse of what the universe would look like if we had not rejected God.

Is God capable of providing us with such glimpses? Yes. Would He have a reason for doing so? Obviously, yes. Such glimpses serve to motivate us towards the new heaven and the new earth that will be brought into existence at the Second Coming.

But there is also another way to look at miracles that doesn't invoke multiple universes at all. A miracle is typically defined as an extraordinary event manifesting divine intervention in human affairs. Thus, we should also try to focus on the intelligence and understanding of the actor who caused the phenomenon.

As Arthur C. Clarke famously wrote, "Any sufficiently advanced technology is indistinguishable from magic." Imagine that we were actually visited by aliens who were able to levitate. We would not automatically assume the aliens were gods who broke the laws of nature. Rather, we would assume those aliens were smarter than us and had figured out how to control gravity or other laws of nature in a way we couldn't, or in a way we didn't understand. Now, if God uses the laws of nature to cause something to happen in an advantegeous way, why do we automatically assume He violated the laws of nature? Again, just because we can't do it, or can't immediately figure out how it was done, doesn't mean it can't have been done.

> The enormous usefulness of mathematics is something bordering on the mysterious . . . . There is no rational explanation for it . . . . The miracle of the appropriateness of the language of mathematics for the formulation of the laws of physics is a wonderful gift which we neither understand nor deserve.
>
> ~ *Eugene Wigner*
> *Physicist and Nobel Laureate*
> *Some hold him to be the equal of Einstein.*

If I lift an object from the floor, I am defying gravity through the flexing of my muscles. In this way of looking at it, a miracle is the name we give to an event in which God flexes His muscles, as it were. Now, God does not have a body, so it isn't as easy to see when He is acting, but His lack of a body doesn't mean He can't act at all.

If I were invisible, a book I carried might seem to violate the laws of nature and fly through the air, even though it were actually obeying physical laws I had manipulated to my own purpose. Similarly, if God were to act, it is reasonable to assume that He would have ways of working within the laws of nature that have not occurred to us. In short, it is no more sinister, spooky or impossible for Him to act in the world than it is for me to act in the world. And why shouldn't He act in the world? It *is* His world, after all.

We have documented examples of miracles, the most well-known being "The Miracle of the Sun" at Fatima. This unusual solar activity was seen by thousands, believers, agnostics and atheists, up to forty miles away. It was reported in the newspapers of the time.

Furthermore, as described in another question, the entire Christian Faith rests on the miracle of the Resurrection. If Christ is not risen, then Paul and all the others are liars, we are still dead in our sins and all of Christianity is a farce.

To deny a miracle on the grounds that God hasn't the right to do such a thing is rather an interesting statement coming from the likes of us. Given the centrality of the miracle of the Resurrection, anyone who denies the possibility of miracles, by that very fact, denies that Christianity is anything but a pack of lies. And that is, in the end, the point of such denials.

11. Will the universe come to an end?

Yes, the universe does come to an end, according to both physicists and theologians. For physicists, how it will end is still not clear. What seems most likely at the moment is a universe populated by nothing but black holes, burnt-out stars and the dead shells of planets. Less likely, but still remotely possible, it might end with everything collapsing back on itself in a Big Crunch, a dimensionless singularity. If the crunch is big enough, this might set off its own Big Bang and create an oscillating universe. There might be an infinite number of universes, each expanding, contracting, crunching, exploding. There are many possible ways for it to come to an end, but physicists do not have enough information to know which will actually occur.

> Is man an unimportant bit of dust on an unimportant planet in an unimportant galaxy somewhere in the vastness of space? No! The necessity to produce life lies at the center of the universe's whole machinery and design.....Slight variations in physical laws such as gravity or electromagnetism would make life impossible.
>
> ~ *John Wheeler,*
> *Professor of Physics, Princeton*
> *Unitarian who contributed to the development of the hydrogen bomb.*

For theologians, the question is both simpler and more complicated. Christ will return with all the saints and angels in glory. When this happens, no one knows, although Isaac Newton felt it could not happen before 2060. On the day of Last Judgement, the furthest consequence of every person's virtue and every person's sin will be made known to every person who ever lived. Everyone will receive their bodies back. That's what the resurrection of the body means – we get our own bodies back, but in a perfected state. In some sense, a new heaven and a new earth will exist, and those who have chosen to live with God will live in glory and peace with Him.

As with the physicists' version of the end of the universe, the details on how all of this happens are pretty sketchy. But those details aren't the point anyway. What matters is that our personal relationship with God will be finalized and perfected.

As we look out into the universe and identify the many accidents of physics and astronomy that have worked to our benefit, it almost seems as if the universe must in some sense have known that we were coming.

~ Freeman Dyson

Theoretical physicist, mathematician and nuclear engineer.
This devout non-denominational Christian made crucial
contributions to both experimental science and pure mathematics.
He is credited with having developed the Dyson Sphere, Dyson Tree,
Dyson Transform and Dyson Series.

Section 2: Evolution

1. Is there a conflict between Genesis and Evolution?

Not necessarily. We must first recognize that we read Scripture according to the literary genres which it presents to us. Books of Scripture like the Gospels, books that are meant to discuss historical events, are read according to a different standard than the Psalms, for instance.

The Catechism of the Catholic Church points out that questions of origin are not something that physical science can answer by itself. Because man is a person, his existence is pregnant with meaning. Physical science can, at most, tell us when a humanoid body appeared. Physical science cannot answer questions that discover "the meaning of such an origin: is the universe governed by chance, blind fate, anonymous necessity, or by a transcendent, intelligent and good Being called 'God'? And if the world does come from God's wisdom and goodness, why is there evil? Where does it come from? Who is responsible for it? Is there any liberation from it?" (CCC 284).

These are the answers Genesis is meant to provide, at least in part. Evolution, on the other hand, is not concerned with any of those questions. Genesis tells us about our relationship with God, not how we got our gallbladder. Indeed, in 1909, the Pontifical Biblical Commission pointed out that we should not expect exact scientific language in Genesis. In 1948, the same commission expounded the problem a little more clearly:

> The question of the literary forms of the first eleven chapters of Genesis is far more obscure and complex. These literary forms

do not correspond to any of our classical categories and cannot be judged in the light of the Greco-Latin or modern literary types. It is therefore impossible to deny or to affirm their historicity as a whole without unduly applying to them norms of a literary type under which they cannot be classed. If it is agreed not to see in these chapters history in the classical and modern sense, it must be admitted also that known scientific facts do not allow a positive solution of all the problems which they present…. To declare *a priori* that these narratives do not contain history in the modern sense of the word might easily be understood to mean that they do not contain history in any sense, whereas they relate in simple and figurative language, adapted to the understanding of mankind at a lower stage of development, the fundamental truths underlying the divine scheme of salvation, as well as a popular description of the origins of the human race and of the Chosen People.

So, Genesis gives us an allegorical, or popular, understanding of a real historical event(s), but it is not a science textbook. It shouldn't be treated as such. Indeed, because it doesn't fit any familiar modern literary form, we should be very careful about how we draw conclusions from it. That's the answer from the side of Scripture. The answer from the side of science is no less nuanced.

We need to remember that there is not just one theory of evolution, but several. For instance, some theories postulate gradual change over the course of time, others require "punctuated equilibrium," that is, long periods with little to no change punctuated by shorter periods with massive change. Just as it is wrong to think of Genesis as being a simple text which can be read either as physical science or straight history, so it is also incorrect to speak of "evolution" as a single or simple concept.

Ultimately, the Church has no problem with evolution so long as the following points are accepted and understood:

- Human beings are defined primarily by the presence of their souls, not the configuration of their bodies,

- All men are related through common descent from a common set of human parents; we are one family. We all share in the original sin of the first parents,

- While the human body may or may not have been subject to evolution, the human soul is not subject to evolution because each human soul is immediately created and infused by God at the moment each human individual is brought into existence. The powers of the soul may mature, but the soul itself does not evolve,

- Whatever biological evolutionary processes did occur are under the guidance of God, at least via his permissive will,

- The truth of evolution is a principle of the biological sciences, it is not a doctrine of the faith. Therefore, Catholics are free to hold to evolution or to refuse it, as they individually feel the evidence warrants,

- No Catholic may question the faith of another for holding a different stance on matters of physical science as long as the above principles are held by all parties. As with any particular teaching of experimental science, this is a subject upon which good Catholics can disagree. However, it is certainly a pious thought to hold that evolution is God's way of continuing to create lower forms of life, if only because we know that conception and birth are God's way of continuing to create human persons.

2. How many theories of evolution are there?

There are several theories of evolution, each with its own nuances and schools. "Phyletic gradualism," for instance, holds that gradual change occurs everywhere over long periods of time. This was Darwin's original view.

"Punctuated equilibrium" holds that long stretches of time see little evolutionary change. These stretches are, however, punctuated by short periods with massive increases in the rate of genetic change in some or all organisms. This causes massive numbers of new species to come into existence.

"Quantum evolution" says small isolated populations develop unusual gene populations that result in the creation of entire taxonomic families.

"Saltationism" says change can arise within a single generation. According to this theory, the child of a sexual union could actually be a different species than its parents.

That's just a short list of the general differences. When specific kinds of organisms are discussed, it can get even more complicated. For instance, the discussion over how to organize single-celled life is enormous. There are at least four major models: (a) the three-domain tree, (b) the eocyte tree, (c) the "web of life" and (d) the "ring of life." All four are mutually exclusive, that is, if one of them is true, the other three cannot be. And that's just the argument over how the simplest, single-celled life-forms came to be what we see today. Once the various discussions involving the various kinds of multi-cellular organisms are added in, the number of theories grows at a staggering rate.

God would not have made the universe as it is unless He intended us to understand it.

~Robert Boyle

Chemist and devout Anglican.
He discovered Boyle's Law in chemistry, helped found the Royal Society and endowed the still-existing Boyle series of lectures, intended to "prove the Christian faith to notorious infidels." He also contributed funds to translate Scripture into several languages.

---

3. Can evolution be proven to be true?

Well, that all depends on what one means by "proven." Given any particular finite set of facts, an almost infinite number of theories can be woven that account for all the facts. This situation is made worse when we are limited to using only the facts, that is, the material, which happened to survive several thousands of years (as with Christianity) or several millions of years (as with archeology) of various insults. When it comes to studying evolution, we don't generally get to choose what we work with. Instead, we tend to get what's left over after fire, flood, famine, plague, insects, bacteria, etc., have given all the physical evidence a good working over.

Thus, believers in both Christianity and evolution face a similar problem. As evolution promoter Bruce Bower points out:

> Given limited evidence about long-gone populations of our predecessors, researchers devise competing evolutionary scenarios that are often difficult to disprove and that can easily accommodate whatever ancient bones turn up next. ("Fossil sparks: new finds ignite controversy over ape and human evolution", *Science News*, Nov., 3, 2007.)

It is no wonder that evolution supporters get as touchy as religious believers when anyone asks probing questions about the science.

There are 36,000 competing Christian denominations, that is, 36,000 competing ideas about how Christianity works. But, if someone appears who denies the divinity of Christ, all 36,000 denominations will attack that idea as absurd. In a very similar way, there are several hundred variations on the precise mechanics of how evolution works, or worked, in various situations. But, anyone who questions the idea of evolution is likewise attacked as having proposed an absurd idea.

The variety of opinion on exactly how to live the Christian life does not prove that Jesus is not God. Similarly, the variety of opinion in evolutionary theory doesn't prove that evolutionists are wrong. People often have vehement disagreements about how to interpret facts, and this holds true whether one is a Christian or an evolutionist or both. The problems facing both are in many ways more similar than either group likes to admit.

4. Why are so many Christians opposed to evolution?

Very few Christians have any real problem with the mechanical processes evolution describes. For instance, it is very hard to deny that environment seriously effects which offspring survive. It is undeniably true that animals with characteristics which fit an environment are more likely to survive than animals that don't have such characteristics. Indeed, the statement is so undeniably true that it is tautological: "The best-suited animals are the ones that survive, and we know those are the best suited because... well... they survived."

Since every experimental science is descriptive, every experimental science is, at some level, tautological. The more reliant the science is on mathematics, the more obvious this will be. Physics and chemistry equations have to balance. Whether we say:

- Black crows are black
- $4 = 4$
- $10 - 6 = 4128/1032$

every one of these statements are tautologies, repetitions, redundancies. If evolution is really a science we can hardly expect it to differ on this essential point. Thus, we should not be surprised to find that even evolutionists admit evolutionary theory is tautology in action. "The fittest survive." Well... yes, quite. Who else is going to survive but the fittest? The less fit, by definition, don't live as long.

In evolution, as in so many other sciences, to make the statement is to assert the conclusion. Evolution, properly done, merely describes what happened. So, no one really argues with the descriptions of the mechanical processes.

The problem rather, is with the philosophy some atheists wish to impose on the scientific theory of evolution. They try to force evolutionary theory into theology, they pretend experimental science can prove God does not exist. Of course, experimental science is completely unqualified to make such an assertion. Real scientists don't try to make it, if only because assertions about God have no bearing on the science involved in describing natural processes.

Yet the assertion is made explicitly, then driven home by insisting that evolutionary processes have "no purpose," that "fitness" does not mean "better" and "mal-adapted" does not mean "worse." The language of evolution often has the faint smell of political correctness, where we dare not declare some life forms "winners" and others "losers" lest the mal-adapted organisms (ignore the fact that "mal" means "bad") lose heart.

For instance, atheists sometimes insist God does not exist, because this or that evolutionary process seems to have been done in a sloppy or inefficient way. Indeed, to our eyes, it is true. But if the engineer specifically designed the building so that it would collapse in a certain way under certain conditions, then one could hardly fault the engineer for having used poor design when the building does collapse exactly as intended.

Science can only describe the mechanics of how a job was done. To decide whether or not the job serves the purposes intended one has to first know the intent of the one who did the job. Atheists insist evolution has no purpose. But how do they know this?

How, for instance, did these atheists who pretend to be scientists test for the existence of "intent" or "purpose"? What device did they use to measure the amount of purpose, or lack thereof, in the particular event(s) they describe? What purpose-filled event did they use as a control to test the accuracy of their "purpose-detection device," what was the direction, force and object of the purpose in the control event? Is the device they used to measure the amount of purpose available for sale? Imagine how popular such a tool would be for teachers, supervisors and everyone else who is tasked with measuring the motivation of those in their charge! You see the problem.

Western religion deals with progress, and progress to some extent suggests purpose. Scientists, on the other hand, try to describe phenomena without invoking purposeful creation.... The theologians seem to have won this time.

~ Arno Penzias

Physicist and Nobel Laureate.
He and Robert Woodrow Wilson discovered the background
radiation which confirmed the Big Bang theory.

Atheism has no predictive power. Even if we grant the premise that there is no God, it does not help us accomplish anything in the world. Our scientific technique does not get better as a result, indeed, many of history's scientists would argue that it gets worse. In fact, the philosophy the atheistic parody of science promotes directly contradicts, it essentially nullifies, our ability to do any science at all.

Consider: if the processes within the universe are purposeless, and if we are a product of purposeless processes, then we are purposeless. But, if everything is purposeless, if none of it means anything, it seems fairly odd that anyone can observe a process, ask the question "I wonder why that happens?" and then investigate it in such a way as to actually obtain a meaningful answer. If nothing has a purpose, why do we do research? Why even ask "why?"

The philosophy of this atheistic version of evolution is rather like the philosophy of the deconstructionist who writes an enormous book to explain why communication between two people is essentially impossible. It may be true, but if the man believed it, why did he write a book? After all, if he believed his own theory, who did he expect would read his words? Or understand them?

In like manner, the atheists who insist evolutionary processes serve no purpose can't explain why science works. Or why we should listen to them if they are right. If the universe really has no purpose, then it doesn't matter whether we teach science in the classroom or not, does it?

You can't have meaning where there is no purpose. At least, no one has successfully explained how to accomplish that. The words "explain" and "accomplish" assume purpose. We have language in order to communicate – we have it for a reason. Thus, the very language we use to discuss the universe necessarily implies purpose, if only because words mean things. Can we use purpose-driven research, described in purpose-driven language, to assert purpose doesn't exist? We can't even ask the question – asking the question assumes a purpose to start and an explanation to come.

So atheistic evolution is, from a philosophical perspective, absolutely absurd, even if biological evolution is, from a mechanical perspective, perfectly accurate.

5. Is there life on other planets?

We don't know. Neither science nor faith takes a position on this question. Science has no physical evidence, although there is a lot of circumstantial reason to think it may, perhaps, exist. From a faith perspective, the answer only matters insofar as there is a moral dimension to the life that is discovered.

For instance, let us assume that we discover incontrovertible proof of bacterial life on another planet. Would that discovery affect Christian beliefs?

No. The presence or absence of bacteria on another planet is not relevant to man's salvation. Similarly, the discovery of plant life or animal life would have no impact on man's understanding of his relationship with God. Even the discovery of persons on another planet or in another galaxy would not have any necessary bearing on how human beings relate to God.

I am convinced of the afterlife, independent of theology. If the world is rationally constructed, there must be an afterlife... My belief is *theistic*, not pantheistic, following Leibniz rather than Spinoza.

Kurt Gödel

Mathematician and Lutheran.
His Incompleteness Theorems revolutionized every logic-based discipline known to man. He used modal logic to demonstrate that Anselm's proof for God's existence is valid.

If we were to discover persons living elsewhere in the universe, or more likely, if they were to discover us, they would exist in one of two possible moral states. Either they would exist as fallen persons, in some way analogous to either the fallen angels or fallen human beings, or they would be unfallen persons, analogous to the unfallen angels or the Blessed Virgin.

In either case, their existence would not alter the facts of our salvation, namely, that we are fallen and that God took on human flesh and died for us.

We are judged by how we treat other persons, period. So if we treat these new and strange persons unjustly or unmercifully, we would be responsible for our failures in justice and mercy, regardless of whether they are fallen or unfallen. Conversely, if they treated us unjustly or unmercifully, we would be responsible for how we responded to this treatment, regardless of whether they can be saved or not.

These strange persons would be responsible before God according to the different aspects of the nature(s) God gave them. We are responsible to God according to our human nature. The Catholic Church is established to save human persons. She isn't established to save angels or any other kinds of persons that might exist. So, the existence of other life, though mighty interesting, simply isn't relevant to Christian theology.

We are responsible for how we interact with the world. What that world consists of, whether bacteria, plants, animals or other persons, is of secondary importance. What matters is this: have we dealt with these other creatures in accordance with the dignity due to them? If so, then we are saved. If not, then we are not.

6. Can science prove the existence of the soul?

Can experimental scientists prove the existence of the soul? No, of course not. Experiment can only test physical things; the soul is spirit, it is not physical. Can the existence of the soul be demonstrated in non-experimental ways? Yes, of course. The existence of this book, your reading it and my writing it, are two bits of evidence. Now, some argue the soul does not exist because scientists cannot demonstrate its existence. But is that fair?

> It was not possible to formulate the laws (of quantum theory) in a fully consistent way without reference to consciousness.
>
> **~ Eugene Wigner**
>
> *Physicist, Nobel Laureate and Lutheran.*
> *He laid the foundation for the structure of the atom and the theory of symmetries in quantum mechanics, along with several important mathematical theorems.*

Things exist which experimental science cannot demonstrate. Memories, for instance. Do you recall an event from your childhood? A conversation, a personal or intimate interaction with someone, perhaps. Whatever historical event you consider, it is not scientifically reproducible, it cannot be verified according to any technique that qualifies as a laboratory test. At best, we might note that certain nerve synapses fire when you recall the event, but that does not demonstrate the existence of the event.

Notice that I am not arguing from a negative. I am not saying, "You cannot prove unicorns do not exist, therefore it is safe to say unicorns may exist." Instead, I am arguing from the reality of your own experiences. You cannot scientifically prove that your own life experiences really took place. You can gather witnesses, you can have them testify to the events, but that is anecdote, not experimental science. It isn't reproducible, much less controlled. Such anecdotal evidence may be good enough for a court of law; it is frequently good enough to get a man jailed for the whole of his natural life; it may even get him hanged. In short, that kind of evidence is more than sufficient for the science of law which judges the interactions between persons, but it is generally not the kind of evidence required by an experimental science which describes the interactions between objects.

To say that experimental science cannot demonstrate the facts of your life experience, that it cannot judge which man should be hanged and which should be merely jailed, that it cannot verify most of the other kinds of assertions involving persons, is to say only that science has to have a different standard of proof because it is dealing with a different kind of knowledge.

If we insist on accepting only what experimental science can prove, then we have greatly impoverished ourselves. To paraphrase, "There is more in heaven and on earth than is dreamt of in our science, Horatio…"

7. How old is the soul and does the human soul evolve?

No, the human soul does not evolve. "Evolution" is a term which refers to how species change over time. Thus, it is a description of parent-child, grandparent-grandchild and great-grandparent-great-grandchild relationships. It describes how these and even longer-chained relationships change over the course of millennia. Evolution describes the chain of physical relationships between multiple generations of physical entities that reproduce.

Every human soul is immediately created and infused by God into the human body at the moment of conception. Indeed, the infusion of a human soul is what makes the body human. It is not the case that one human soul begets or conceives a child soul – each soul is created specifically by God for the human body that will receive it.

> When I reflect on so many profoundly marvellous things that persons have grasped, sought, and done I recognize even more clearly that human intelligence is a work of God, and one of the most excellent.
>
> ### Galileo Galilei
>
> *Physicist, mathematician, astronomer and devout Catholic.*
> *He made major contributions in hydrology, dynamics, astronomy*
> *and mechanics. Many consider him the Father of Modern Science.*

Human parents conceive the physical body of their children, but they do not participate in the creation of the human souls of their children, except through the indirect action of providing the body into which the soul is infused. Because there is no "parent-child" relationship between human souls, there can be no evolution of human souls.

In the same way, each human soul is as old as the person who is that soul. The soul is the form of the body, it shapes the body into what we see. The human soul does not exist from all eternity to all eternity, nor does it pre-exist the human body that it molds and forms. The human body and the human soul come into existence together, the body is supplied by the parents through sexual union, the soul is supplied by God through an immediate act of creation. Thus, the body is subject to evolution, the soul is not.

Because the soul is not physical, it is not detectable by physical science. We can see its effects on the world, and in that way gain evidence of its presence or absence. So, an hydatidiform mole is known to be without a human soul precisely because the entity does not form itself into a human body. Similarly, what looks to our eyes and our tests to be a single body and therefore a single soul may, in fact, turn out to be two souls in what turns out to be two bodies joined together as one, as in the case of Siamese twins. Although these two persons may have been joined together from the first moment of their lives, they are and have always been, from the moment of conception, two distinct persons, each with his/her own soul.

8. Did we descend from monkeys?

Our souls do not come from animals. But, over the course of time, our bodies were certainly subject to factors that favored the survival of some characteristics over others, e.g., different skin colors, differing ability to digest milk or metabolize alcohol, etc. For this reason, we need to be very careful in distinguishing the soul and the body.

The human soul is the form of the body. That means the human soul molds, shapes, affects the growth and appearance of the human body. The soul is affected by grace. Baptism, Confirmation and Holy Orders are each sacraments which put a stamp or seal upon our souls.

The unbaptized soul is substantially different from the baptized soul. But, when the waters of baptism or the oil of Confirmation or Holy Orders anoints the person, the finger of God reaches through those outward signs and actually changes who we are. My baptism transforms me into a new creature. I am no longer just a creature of God; through baptism, I am changed into a child of God. Science has no way of determining through experiment who is baptized and who is not, but that is not relevant. The change is there.

In a similar way, the union of the human soul to a body is a large part of what makes that body human. This is why we can say a human zygote or human embryo is a human person even though the body we observe only looks like a single cell or a collection of cells, and nothing like a child or an adult. The cell is human. The soul is human. A body that has been united to a human soul is a human body, even when the soul departs, as it does at death. Archeological and experimental sciences have no way of determining if the dry bones they study were once enwrapped by a human soul.

The evidence supporting the idea that all living things are descended from a common ancestor is truly overwhelming. I would not necessarily wish that to be so, as a Bible-believing Christian. But it is so. It does not serve faith well to try to deny that.

Francis Collins

Geneticist and self-described Bible-Believing Christian.
Director of the Human Genome Project. The Endocrine Society
described him as "one of the most accomplished scientists of our time."

When we look simply at physical bodies - at DNA, RNA, mitochondria and other cellular structures - we cannot necessarily tell what species is before us. True, every set of human DNA is 99% identical to every other and human parents are 99.5% identical to their children. But we must remember that chimpanzees share almost 99% of their DNA with us, orangutans share 95%, even plants share upwards of 50-60% of their DNA with us. So, depending on which stretch of DNA you look at, you could conclude that I am really a sunflower, or that a fruit fly is eligible for Social Security. For certain stretches of DNA, there is no experimental test that could tell us whether it was part of a plant, an animal or a man.

So, it is quite possible that there once existed hominids which looked very similar in their body or even in their DNA. But that is not the whole story. If those hominids did not possess human souls, they were not human, anymore than a banana is human simply because it has many of the same DNA sequences.

Human beings are far more similar than we realize. Even though chimpanzees are almost 99% identical, there is more variability between two chimpanzees from neighboring clans than there is between two human beings from opposite sides of the earth. That *one percent* difference is not only an infinite chasm between man and chimp, it is a chasm which doesn't separate any two men. "Even today, we are a species of 6 billion with the genetic diversity of a population of a few thousand." (Bruce Bridgeman, *Psychology and Evolution*, Sage Publications, 2003, pp. 55-56). Scripture tells us that we human beings are all one family. Molecular genetics strikingly confirms that idea.

So, just as God changes us through the sacraments, it is possible that at some point, God reached down and changed one of these hominids - ensouled one of them, breathed human life into him. Scripture says, quite beautifully, that God made man out of the clay of the earth. It doesn't say exactly how He did that, nor how long He took to get it done. Could He have formed the body through evolution? The 1910 writers of the famous Fundamentals theological tracts agreed that He could.

Does this mean we descended from monkeys? Well, we must first leave aside the fact that modern monkeys are not the same as the common predecessor from which both of our bodies may have evolved. But, even leaving that aside, the question, when phrased in a theological context, doesn't make very much sense. There is no way for experimental science to affirm or deny the statement, since it has no test to determine whether those old dry bones it studies were ever wrapped around with a human soul. We can note lots of body similarities, even DNA similarities, but many of these physical similarities are common to all life.

Are our bodies biologically related to modern monkeys? Undeniably, they are. Did our human souls come from monkeys? Clearly not. God descended from the heavens for us, and we came from His hand, in whatever way He chose to make it happen.

9. If we are God's co-workers and evolution is God's work, isn't it reasonable for us to help evolution along by getting rid of inferiors? Shouldn't we try to breed a race of thoroughbreds?

This idea is called eugenics, and it was commonly accepted as cutting edge science from the late 1800's through the mid-20th century. Eugenics sprang from interpretations of Charles Darwin's evolutionary theory. While Darwin himself never explicitly endorsed eugenics, he certainly had some sympathy to the idea. The following quote is merely a sample of his conflicted thoughts on the subject:

> With savages, the weak in body or mind are soon eliminated; and those that survive commonly exhibit a vigorous state of health. We civilized men, on the other hand, do our utmost to check the process of elimination; we build asylums for the imbecile, the maimed, and the sick; we institute poor-laws; and our medical men exert their utmost skill to save the life of every one to the last moment. There is reason to believe that vaccination has preserved thousands, who from a weak constitution would formerly have succumbed to small-pox. Thus the weak members of civilized societies propagate their kind. No one who has attended to the breeding of domestic animals will doubt that this must be highly injurious to the race of man. It is surprising how soon a want of care, or care wrongly directed, leads to the degeneration of a domestic race; but excepting in the case of man himself, hardly any one is so ignorant as to allow his worst animals to breed.
>
> The aid which we feel impelled to give to the helpless is mainly an incidental result of the instinct of sympathy, which was originally acquired as part of the social instincts, but subsequently rendered... more tender and more widely diffused. Nor could we check

our sympathy, even at the urging of hard reason, without deterioration in the noblest part of our nature. The surgeon may harden himself whilst performing an operation, for he knows that he is acting for the good of his patient; but if we were intentionally to neglect the weak and helpless, it could only be for a contingent benefit, with an overwhelming present evil.

Hence we must bear without complaining the undoubtedly bad effects of the weak surviving and propagating their kind; but there appears to be at least one check in steady action, namely the weaker and inferior members of society not marrying so freely as the sound; and this check might be indefinitely increased, though this is more to be hoped for than expected, by the weak in body or mind refraining from marriage. (Charles Darwin, *The Descent of Man*, Chapter 5, "Natural Selection as Affecting Civilised Nations")

As you can see, while he admires tenderness, he constantly laments the "bad effects" of inferior breeding. Darwin's eugenics tendencies influenced many people. Every American president from Theodore Roosevelt through FDR was an avowed eugenicist, and so were most highly educated Americans and Europeans. American scientists from Stanford, Yale, Harvard, and Princeton persuaded millionaires like Harriman, Rockefeller and Carnegie to set up foundations to promote eugenics, both here and abroad. Indeed, the earliest eugenics experiments in Germany were funded by the Rockefeller Foundation, not the Nazis. Those foundations exist today.

The first involuntary sterilizations ever legally performed on adults in the modern era took place in Indiana (1911), although they had been performed illegally in Indiana prisons for years before. An Englishman, Francis Galton (Darwin's half-cousin), endowed a chair of eugenics studies at the University of London upon his death in 1911. Eugenics journals were common throughout Europe and the United States. The labs at Cold Spring Harbor, New York and several locations in California were the centers for American eugenics activity.

> [F]ew indeed, will admit the reality that God made man in His image in which case all earth men are alike. There is in fact but one race, of many colours. Christ is but one person, yet he is of all people, so why do some people think themselves better than some other people?
>
> **~Nikola Tesla**
>
> *Serbian Orthodox mechanical and electrical engineer.*
> *He invented the AC system of power generation and transmission.*
> *The Standard International unit of flux density is named for him.*

But, though Englishmen invented the "science" of eugenics and Americans were the first to implement it, by the 1930s, leading intellectuals like Margaret Sanger were concerned that Nazi Germany was leapfrogging ahead. Germany, after all, modeled the Nuremburg code regulating marriage between Aryans and Jews on Virginia's law forbidding intermarriage between blacks and whites.

During the war, Germany took those laws to their logical conclusion. Still, as more than one defendant pointed out during the war crime trials after World War II, the Nazi law was actually more lenient in who could be married than US law was, since the Nazis considered anyone with 1/8th part Jewish ancestry to be Aryan, while US law insisted anyone with more than 1/32nd part black ancestry was not white.

The 1927 US Supreme Court ruling *Buck vs. Bell* established that the state has a right to sterilize anyone. About 64,000 Americans were forcibly sterilized between 1907 and 1974. *Buck vs. Bell* has never been overturned. As of March, 2009, at least sixteen American states still have no laws on the books protecting citizens from being subject to involuntary sterilization.

Today, educated people still carry sympathies towards eugenics. In 2006, Richard Dawkins, the Charles Simonyi Professor of the Public Understanding of Science at Oxford University, wrote the Scotland Sunday Herald and asked what the moral difference is "between breeding for musical ability and forcing a child to take music lessons. Or why it is acceptable to train fast runners and high jumpers but not to breed them. I can think of some answers, and they are good ones, which would probably end up persuading me. But hasn't the time come when we should stop being frightened even to put the question?" Gregory Pence pursues the same line of thought in *Who's Afraid of Human Cloning?*, when he asks, "Would it be so terrible to allow parents to at least aim for a certain type, in the same way that great breeders... try to match a breed of dog to the needs of a family?" (p. 168)

Today, eugenics is considered junk science by all but the non-Christians and atheists, who have always promoted it. Throughout the decades when eugenics was considered enlightened cutting-edge science, its supporters always wailed about one powerful, determined opponent: the Catholic Church. As eugenics gains ground again, atheism is again attacking the Church.

Religion understands what experimental science does not: every person has value, every person must be treated with the dignity of a person created in God's image and likeness. Children are not animals to be bred, but gifts from God who are welcomed with joy. Involuntary sterilization is an offense that insults both the one who is sterilized and the one who sterilizes. Trying to breed men as if they were dogs is an offense against man and God.

Throughout the late 1800's and the early 1900's, the Church constantly exhorted nations to avoid this sin, for it would certainly lead to greater evil. Her most strenuous pronouncement was the 1937 encyclical *Mit Brennender Sorge*, in which Pope Pius XI specifically condemned "substitutes or arbitrary alternatives such as certain leaders pretend to draw from the so-called myth of race and blood."

Human beings do not breed. Animals breed. Human beings procreate – we participate in the divine creating power of God. We help to bring into existence immortal human persons, persons who will exist for all eternity. Our lives on earth are but dust on the scales compared to the length of life we can spend with God.

10. If eugenics is bad, what kind of genetic tinkering is moral, what kind of genetic tinkering is immoral?

Genetic manipulation is not, by itself, immoral. The use to which it is put is what affects the morality of the technology. For instance, the use of DNA manipulation in animals for testing, treatment or the development of new therapies is perfectly reasonable.

Normally, any treatment that serves to heal a human being without simultaneously putting other human beings at risk of serious injury or death is perfectly acceptable. But when it comes to DNA regulation, more than usual care must be taken. Our DNA is not just ours. It is a patrimony handed down from parents to child through countless generations.

In each generation the parents contribute their own bodies. Down through the ages, each parent gives half of the blueprint for his/her body to his/her children. In each generation, each child receives this accumulated gift from parents, grandparents, great-grandparents, all who have gone before them.

> The knowledge that we get about human biology and human genetics is neither good nor evil. It's just knowledge. The application that we choose for that knowledge can take on moral character.
>
> **Francis Collins**
>
> *Geneticist and self-described Bible-Believing Christian.*
> *Director of the Human Genome Project. The Endocrine Society*
> *described him as "one of the most accomplished scientists of our time."*

But at the same time, as described above, we share a lot of DNA with animals; almost 99% with the chimpanzee, for instance. Indeed, some of the animal tissues produced by the DNA found in animals are close enough to human tissue to permit direct transplant. For example, heart valves from pigs, horses and cows are commonly used for heart repair in human beings. This kind of transplantation from animals into humans is not immoral. So, using a stretch of DNA from a human being which is identical to a stretch of DNA in a chimpanizee or a banana would not necessarily create a moral problem.

However, because DNA and RNA manipulation has such a great potential to significantly alter the human genome if care is not taken, it is immoral to work in this area without proper safeguards to protect against this possibility. This need to safeguard the human person is particularly important when using viral vectors that might alter human DNA in unforeseen ways, especially if there is a risk of altering germ cells (sperm and ova).

But genetic manipulation concerns don't stop with the actual manipulation of the genome. Consider, for instance, the farmer who grows food. It is possible to use genetic manipulation produce a highly sought-after crop which cannot reproduce. When the seeds are bought and sown, the resulting crop does not produce seeds, or produces only sterile seeds. In other cases, the seeds are fruitful, but are embargoed by legal contract so that the farmer cannot use them or sell them the next season. He must throw away what he cannot sell and buy new seed.

While it is understandable that the men and women who poured millions into the research for a new hardy and/ or fruitful crop deserve just payment for their work, it is also understandable that the farmer who uses such a crop would need time to adjust his expectations and outlook on what he can do with it. In short, it is not clear what effect genetic technology has on economic justice.

Moral issues would also arise when dealing with the one-percent of DNA which is unique to the human person.

It is now known that sequences in this one percent affect brain form and function, hand formation, and facial muscle and vocal muscle control. Other sequences, although unique to humans, are not necessarily properly human at all. They are actually the remnants of retroviruses which managed to embed themselves into the human genome at some point in the past. From a theological perspective, should these sequences be considered human or not? We must think more deeply on this question.

Genetic therapy that heals a disease or disfigurement in an existing person is morally acceptable. But gene manipulation that can affect future generations, that is, interventions that affects the germ line, the egg and/or the sperm, is not acceptable. If there is a question concerning whether a manipulation would affect the germ line, then it could not be done.

For instance, we could not manipulate the genetics in an embryo prior to the differentiation of the gonads because any change made to the embryo would necessarily ripple out into the future gonad tissue. Genetic manipulation would have to wait until we were assured that future children would not be affected. It is not our place to redefine the human body for future generations who might not want the kind of redefinition we are imposing.

11. What is the difference between technique, theory and philosophy and how does this affect evolution?

Scientific technique is the act of observing, measuring and recording data. Scientific theory attempts to explain what we observed and measured, in order to predict what we should be able to observe and measure under other circumstances. Scientific philosophy is the system for making sense of why sciences techniques and theories work. Some philosophies of science are very simple, very mundane, they restrict themselves to finding cause and effect on the physical level. Other philosophies look for a "meta-physic", a larger, uniting vision. Strictly speaking, philosophy and metaphysics are not experimental science.

Although philosophy is not the same as science, the philosophy a scientist holds will affect the theories he forms and may even affect his observations, what he chooses to measure and record. Obviously, we only record what we think is important. If two scientists hold differing philosophies, what they think is important - what they bother to record - may be very different indeed.

> I see a universe that, had it been constructed slightly differently, would never have given birth to stars and planets, let alone bacteria and people. And there is no good scientific reason for why the universe should not have been different. Many good scientists have concluded from these observations that an intelligent God must have chosen to create the universe with such beautiful, simple, and life-giving properties.
>
> **William D. Phillips**
>
> *Physicist, Nobel Laureate and devout Methodist.*
> *He won his Nobel Prize for developing the technique of laser cooling.*

For instance, as we saw in the Cosmos section, astrophysicist Fred Hoyle developed a Steady State theory of cosmology not because the evidence demanded it, but because he felt his personal atheism was at odds with the philosophy which guided the Big Bang idea. He wanted an alternative. He developed an alternative theory which accounted for all the evidence, but did not account for later discoveries like the cosmic background radiation. He eventually discarded his atheism.

Meanwhile the men who discovered the background radiation did not expect to find it. At first, all they knew was they had a radio hiss in their receivers that really threw off their attempted observations. They didn't try to measure it or observe it. Instead, they tried to get rid of it in every way they knew. When nothing worked, when they were forced to face the reality that they simply couldn't get rid of it, they changed their philosophy. They realized that this was not unimportant noise, instead, this background hiss was actually the important observation. Once their philosophy changed, they were able to see that they had found the cosmic background radiation which proved the Big Bang.

Now, if they had gone into the test expecting to find the background hiss, their techniques for dealing with it would not have been any different. They still would have had to try and "get rid" of the noise in order to prove that the noise wasn't due to any man-made or natural interference, but they would have had a different attitude as they tried to accomplish it. In this case, their philosophy didn't influence their technique, but it did influence their observations of what was important and what wasn't.

Faith in our philosophy plays an important role in how we search for evidence. For instance, in cosmology, the idea of the Grand Unification Theory (GUT), that is, the idea that all cosmological forces can be brought together in a single equation or set of equations, is a philosophical statement. We don't know how to bring all of these forces into a single equation, but we have a faith-filled belief that it can be done. The evidence which supports this faith is the fact that electricity and magnetism have already been united in this way, as has the nuclear weak force. We haven't been able to incorporate the nuclear strong force or gravity into the mix, but – even though we cannot currently imagine how it can be done - we believe that it can be done.

This is faith. Faith is the evidence of things not seen. We have not seen a Grand Unification Theory, but we believe in it because we have seen prior evidence which points to the possibility – the electromagnetic-weak synthesis. But the only reason we think this counts as prior evidence "pointing to" the possibility of a GUT is that we have faith in the idea of a rational universe.

How do all of these concepts affect evolution? It doesn't necessarily affect what we do in terms of technique, of measurement and observation. But it may affect what we think is important in our observations. Evolutionary technique measures the differences between species' body shape, DNA, habitat and other variables. Evolutionary theories attempt to explain why we see the different species we do. Evolutionary philosophies or metaphysics attempts to create a grand unifying vision that encompasses all the different archeological and living evidence.

12. Is Intelligent Design science?

Intelligent Design is a philosophical approach to doing science, the kind of philosophy we discussed above. It is not a method of measurement or observation, so it is not a branch of experimental science. Rather, it is a philosophy which seeks to bring meaning to measurements and observation.

But, though Intelligent Design may not be experimental science, neither is it junk. As we already noted, every set of observations is filtered through a specific philosophical framework. Great scientific work is predicated on good philosophical underpinnings. Philosophy helps us make sense of our observations. From the time of Galileo until today, science has been built on Intelligent Design, the idea that the universe is knowable, rational and worthy of investigation, that it is created and guided by some universal intelligence. Throughout its history, Intelligent Design has been a remarkably successful philosophy for organizing and directing observation and measurement.

Believing that the power to destroy belongs to the Creator alone I affirm... that any theory which, when carried out, demands the annihilation of force, is necessarily erroneous.

~ *James Prescott Joule*

Physicist and Catholic.
His work on thermodynamics and the law of energy conservation led to the Standard International unit of energy named for him.

To give just one example, as we saw previously, astrophysicist Fred Hoyle developed a Steady State theory of cosmology not because the evidence demanded it, but because his personal philosophy, atheism, did. He eventually discarded his atheism and, like many cosmologists, he now promotes an Intelligent Design concept of cosmology. He insists that "A common sense interpretation of the facts suggests that a superintellect has monkeyed with physics, as well as with chemistry and biology, and that there are no blind forces worth speaking about in nature."

Note that these comments come from an astrophysicist, a mathematician of a very high order. Astrophysics is just about as "hard" as science gets, yet numerous physicists and astrophysicists are either theists or converted to theism as a direct result of their work.

Biology, on the other hand, is a much "softer" science, much less driven by hard numbers. Indeed, some members of the "hard science" community don't consider biologists part of the group. At the 2005 Templeton Cambridge Journalism Fellowship lectures, physicist and theist John Barrow derided the atheist Richard Dawkins by laughing at him and saying, "You have a problem with these ideas, Richard, because you're not really a scientist. You're a biologist."

Many proponents of evolutionary philosophy insist that evolutionary science techniques of measurement and observation shows an essentially meaningless, purposeless existence. This insistence is, like Intelligent Design, simply an expression of a philosophy. It is not itself science. To be honest, it is also pretty lousy philosophy.

After all, a scientists should know that science has no technique, no method, for detecting purpose or meaning. There is no lab instrument which is sensitive enough to determine what someone or something means. Science technique can only observe what happened and, perhaps, record the observations. To say that the scientific method used in biology does not detect purpose, therefore there must not be any purpose, is absurd. It is an argument from silence, the silence of a method that everyone agrees isn't capable of measuring the very thing that is being denied.

Only philosophy can make statements about meaning, the why and the wherefore. So, when a scientist insists that something either (a) has meaning or (b) does not have meaning, he is no longer doing science, he is doing philosophy. It doesn't matter if he is wearing his special white lab coat when he makes the statement. Neither (a) nor (b) is empirically testable, neither is subject to any kind of experimental scientific verification. Both statements can only be investigated through theological means.

When I wrote my treatise about our [solar] system, I had my eye upon such principles as might work with considering men for the belief of a deity; and nothing can rejoice me more than to find it useful for that purpose.

~Isaac Newton

Physicist, mathematician, astronomer.
Possibly the most influential man in all scientific history, he made advances in optics, light, color, physics, mechanics, gravitation and math. His Principia Mathematica, in which he describes the calculus he invented, has a section devoted to extolling the Divine Creator of the Universe using an Intelligent Design argument. Though he made more advances in science than most men ever have, he wrote more theology in his lifetime than he did science.

Section 3: Ethics

1. When Did Life Begin?

Neither science nor theology gives us a specific date for when life began. Incidentally, even if one of the theories of evolution is correct, it only tells us how living things change over time. Strictly speaking, the theories of evolution are not able to indicate when life began or how that occurred. That is beyond the scope of the idea.

We don't know how inanimate objects became animate beings. We don't even have a good model for how that change occurred.

Biology has a certain schizophrenia in this regard. The theory of spontaneous generation, the idea that living organisms arise from non-living materials, is a condemned proposition in biology, proven to be wrong by no less then Louis Pasteur, the devout Catholic who helped build the foundations of microbiology. Because of the work of this man, who prayed a daily rosary, we know that it is not the case that bacteria spontaneously form in a broth, or that straw spontaneously generates mice. I can't catch a cold simply by being in a freezer or a snowstorm – I must be infected with the virus first.

However, in order to get life out of inanimate objects, biology must assume spontaneous generation happened at least once. Biology does not explain why it happens only once, or why spontaneous generation does not happen over and over again during the course of millennia.

But the question, "When does life begin?" is sometimes asked in reference to human life. At what point does an individual's human life begin? In that context, the question requires a careful answer. After all, every cell of the human body is alive. Egg cells and sperm cells are alive. The fertilized egg that the union of egg and sperm produce is alive. It is incorrect to say that life begins with the fertilized egg, since everything prior to fertilization was already alive. So, if this question is asked in reference to the origin of an individual human life, it is better to ask, "when does the individual person begin?"

2. What is the distinction between life and human life?

If we understand the soul to be the life force, or organizational principle in a living thing, then every living thing has a soul: grass, trees, ants, bees, dogs, deer, etc. When any one of these living things die, its organizational principle, its life force, is destroyed.

Human beings are distinguished from animals in one major respect. Our soul, our life force, our organizing principle, is pure spirit. Spirit is immortal – it cannot die. God is pure spirit and angels are pure spirit. Man is a body enlivened, ensouled by a spirit. Human beings are unique. We alone have "spirit-souls." The human soul is made up of the human intellect and the human will.

It is our rationality, known through our intellect, expressed through our will, that makes us in the image and likeness of God. Our souls are different than the souls of any other animate creature, our souls are immortal spirits. Each human being has a unique human intellect and human will. Only human beings have an immortal spirit as an organizing principle for their bodies.

A person is that which possesses an intellect and a will, personhood is the most perfect way of being. God is Three Persons, every angel is a person. Of all the creatures with physical bodies, only human beings are persons.

3. When does personhood begin?

A human person consists of a human soul (the human intellect and the human will) joined in union with a human body. Thus, personhood begins when the human soul is infused into the human sperm and the egg as they unite to form the fertilized egg. Embryologists all agree that the individual human life begins with the union of egg and sperm. It is interesting to notice that, of all the kinds of scientists who have been asked to give testimony before courts and Congress, almost none have been embryologists. The beginning of every human life is a scientifically verifiable moment.

> The reason I went into this field [genetics] was to figure out how to treat illnesses, rather than try to stop such individuals from even being born.
>
> ### *Francis Collins*
> *Geneticist and self-described Bible-Believing Christian.*
> *Director of the Human Genome Project, the Endocrine Society described him as "one of the most accomplished scientists of our time."*

Keep in mind that the human egg is generally fertilized in the upper one-third of the fallopian tube. From the moment of fertilization, the egg begins the process of cell growth and division. Within hours of the original union, the first cell division takes place. As soon as the entity is multi-cellular – containing two or more cells – the child is no longer a fertilized egg, but an embryo. The embryonic child must then travel down through the remaining two-thirds of the fallopian tube and implant in the uterine wall in order to continue growing out the human body the child will need in fully functioning form at birth.

Thus, despite what many medical personnel say, it is never the case that a fertilized egg implants in the womb. Indeed, it is impossible for the child to implant in the womb at the single-cell stage. Only an embryo can do this, and the body of the embryonic child must grow substantially in order to successfully implant. Upon implanting, a process which normally takes about six days, the child creates his or her own placenta and establishes the biological relationship within the mother which completes the pregnancy begun at conception.

Pregnancy and personhood begin simultaneously at fertilization. True, the existence of the embryo only becomes apparent to the mother and to the outside world at implantation. It is only at implantation that the embryo begins to emit the hormonal signals into the mother's bloodstream that allow us to detect its presence. However, it is incorrect to say pregnancy begins at implantation since the embryo exists within the woman before implantation.

4. Is science capable of being ethical?

First, we have to distinguish between "ethics" and "morality." Ethics describes how we conform ourselves to a particular set of man-made rules. Morality describes how we conform ourselves to our own human nature, that is, how to live in accord with the divine law inscribed in our heart. Ethics corresponds to what we do, morality corresponds to who we are. Now, because we are material creatures, what we do molds who we are to a certain degree, so there is a certain amount of overlap, but the two can be distinguished.

> Believe in God, in His providence, in a future life, in the recompense of the good; in the punishment of the wicked; in the sublimity and truth of the doctrines of Christ, in a revelation of this doctrine by a special divine inspiration for the salvation of the human race.
>
> ### André-Marie Ampère
>
> *Physicist, mathemetician and devout Catholic.*
> *One of the key explorers in electromagnetism, the Standard International unit of electric current is named in his honor.*

Consider: it might be unethical (that is, a breach of a particular code) for a doctor to sleep with one of his patients; but it is not necessarily immoral. If the doctor's patient was his wife, then it would be perfectly moral for the doctor to sleep with this particular patient, although it might be unethical of him to accept his wife as his patient. On the other hand, there might be nothing in the medical code of conduct for this particular medical specialty (podiatry, for instance) which forbids doctors from sleeping with their patients. So it might be ethical for a doctor to sleep with one of his unmarried patients, but it would be immoral.

So, science is ethical, that is, scientists are capable of drawing up and adhering to various codes of conduct. But these codes of conduct are not necessarily moral.

The medical code of conduct during war permits triage: the treatment of the wounded according to the severity of their wounds. Thus, a slightly wounded patient might be put off for a time in order to treat a more severely wounded patient, while a very severely wounded patient might receive little or no treatment at all, because the amount of resources (time, medications, skill levels) are too limited to save the patient's life. All we can do is help him be comfortable as he dies because we cannot save him.

This concept of triage is both ethical and moral. But, as Robert Jay Lifton pointed out in his book *The Nazi Doctors*, this same triage concept was used by medical doctors in the National Socialist (Nazi) death camps to justify the gassing of millions. "Which is better," a doctor new to the camp would be asked, "Should we try to keep small children, the old, the sick alive in this hell-hole, or give them a clean and easy death so they don't have to suffer? We don't have the resources to save them all. These small and weak ones will die no matter what we do. Don't we give our own troops triage? Then why would we refuse it to the people arriving in the cattle cars? Better they go to heaven in a cloud of gas than suffer starvation and disease here."

The camp conditions were under the control of the very same government that sent people to the camps. The people were better off where they had been. But, since the government saw these people as threats, they deliberately placed them in a highly dangerous situation so that they would be killed. King David used a similar tactic to deal with an inconvenient husband.

Thus, medical professionals at Auschwitz, all licensed doctors, many well-known and well-respected in their fields, were acting ethically, that is, they acted according to a particular code of conduct, but what they did was immoral - it is wrong to kill someone just to end suffering. Their medical work adhered to a certain set of human rules (kill the unhealthy because we don't have the resources to allow them to live), but that code of conduct violated everyone's humanity, both the humanity of the doctors carrying it out and the humanity of the people upon whom it was carried out.

Insofar as science recognizes the existence of persons and treats those persons with the dignity due them, science is both ethical and moral. If science fails to recognize the persons upon whom it operates, or confuses non-persons with persons, then it may operate ethically, but its actions will certainly be immoral.

Since science is incapable of identifying the presence of the human soul through experiment, experimental science cannot operate both ethically and morally unless it is informed by theology.

5. Is it ever justifiable to use weapons of violence?

Yes. There are numerous examples of good men doing just this. Moses, Joshua and David all led armies against their enemies. Jesus whipped the money-changers out of the Temple (John 2), He said He came not to bring peace, but a sword and division, even within families (Matthew 10:34). However, Christians may only use force of arms to protect the weak, the innocent and those in need. Christians have a duty to protect innocent life, even if the innocent life they protect is their own. As Aquinas points out:

> If a man in self-defense uses more than necessary violence, it will be unlawful: whereas if he repels force with moderation, his defense will be lawful. . . . Nor is it necessary for salvation that a man omit the act of moderate self-defense to avoid killing the other man, since one is bound to take more care of one's own life than of another's. (*Summa Theologica*, II-II, Q64, A7)

This duty to use force to protect yourself and others is especially grave when the lives of others are your responsibility (CCC 2265). This is why the state, which is charged with the protection of the lives of its citizens, has a right to engage in capital punishment and just war. The criteria for both are essentially the same (from CCC 2309):

- the damage inflicted by the aggressor on the nation or community of nations must be lasting, grave, and certain;
- all other means of putting an end to it must have been shown to be impractical or ineffective;
- there must be serious prospects of success;
- the use of arms must not produce evils and disorders graver than the evil to be eliminated.

While many people decry religion as the main cause of violence in human history, this is actually not true. Science has created far more war casualties and deaths than all the wars of religion combined. The Franco-Prussian war, World War I, World War II, the Korean and Vietnam wars were all wars fought over the sciences of economics and/ or eugenics.

While most wars prior to the 1700's were fought over religion, one can hardly blame the men of that age for invoking Christ. Who or what else would they invoke? Experimental science did not really begin until Galileo, Europe had no unifying philosophy other than Christianity, public action could only be justified through public appeals to religion. However, as other kinds of science developed – economics, biology, chemistry, sociology, etc. – men turned to these sciences to justify violence and warfare. Christian religion dropped out of vogue.

The reason was simple: Christianity was never a very good tool for war, and it was discarded as soon as better philosophical tools came to hand. The best excuses to wage war are found in the non-theological sciences. Indeed, experimental science is uniquely suited to warfare since it not only provides the philosophical justification (economics, biology, etc.), it also provides wonderful advances in creating weapons of mass destruction.

6. Is science the only path to truth?

God is truth and the source of all truth. Christ is the only way to come to knowledge of truth.

However, all that is made came into existence by, through and for Christ. So, the study of the natural world is one way of coming to knowledge of the truth because the natural world comes to us through Christ.

Obviously, study of the natural world is not, by itself, sufficient to fully understand the truth. Consider, for instance, how you get to know a person. First, you become aware of their reality – you see some trace of them in the universe that you observe. Perhaps the person is in the room with you. That is the fullest kind of natural revelation. From this kind of natural revelation you can glean quite a lot of scientific information: height, weight, hair color, girth, many of the person's physical characteristics will be apparent.

Of course, many physical characteristics won't be apparent. You won't necessarily know if the person standing in the room with you has high blood pressure, cancer, bad kidneys, unusual strength. That would require further scientific testing.

If the person is not within sight, you may not get that much physical information. Perhaps you only know of this person because they are a pen pal, or you both comment on the same blogs. In that case, you only know them by trace; a testimony in pen and ink, the evidence of glowing phosphor or proof by light-emitting diode.

How do you get to know more about this person? Well, they have to invite you, they have to reveal themselves. I can X-ray, cat scan and MRI someone all day long, but never find out if they like Mozart or if they think the Marx brothers are funny. Personal revelations like that can only come from personal communications. Ironically, then, I may know a person better because of their traces, their letters or e-mails, then I know them through physical interaction, if only because they have more fully revealed themselves to me through their personal communications.

How many people in your life do you physically interact with every day, yet you barely know their name? How many others do you not physically see for months at a stretch, but you know them almost as well as you know yourself? This is, in a certain sense, the difference between a "scientific" knowledge and a "theological" knowledge. The person we see every day but do not know, that is a scientific knowledge. The person we never see, but know intimately, that is theological knowledge.

There is no requirement that every statement be a scientific statement. Nor are non-scientific statements worthless or irrational simply because they are not scientific. "She sings beautifully." "He is a good man." "I love you." These are all non-scientific statements that can be of great value. Science is not the only useful way of looking at life.

William D. Phillips

Physicist and Nobel Laureate
This devout Methodist won his Nobel Prize for developing the technique of laser cooling.

From this example, we can see that the simplest form of truth is natural revelation. "The heavens are telling the glory of God." Physical revelation is good and useful, but only to a degree. Precisely because nature is mostly made up of objects, precisely because nature is not a person, the natural universe cannot communicate very much about the person of God. Study of the universe can give us traces, sometimes very interesting traces, but nothing more.

Only a person can reveal himself or herself. The divine Persons of God must reveal Himself to us or we can't know Him intimately. Further, we will only know as much as He is willing to let us know. This kind of interpersonal revelation between the Persons of God and the persons of mankind is called divine revelation.

Divine revelation relies on natural revelation. The prophets speak with their mouths, move wind around to form words, push pen on paper to create documents. Ultimately, however, divine revelation is more than just the natural world. It is the Divine Persons giving us a specific message or set of messages about Himself.

In days past, a man interested in a woman would secure the services of a matchmaker, who would plead his case before her. The matchmaker revealed to the woman the kind of person the man was, but ultimately, the matchmaker could only do so much. At some point, the matchmaker has to step aside and let the man talk for himself.

The prophets were matchmakers, persons who revealed much more than the physical qualities of the natural world could reveal, but persons who ultimately had to step aside. At some point, God has to speak for Himself.

That is why He took flesh and walked among us as a man. He pled His own case, revealing Himself fully so that we could enter into intimate relationship with Him. This is why Jesus is the only way to come to intimate knowledge of the Truth. God is Truth. Truth is Persons.

7. Can morality exist apart from religious belief?

No. Morality is the proper expression of our human nature. When we act morally, we act in accord with the fact that each of us is an image and likeness of God. In order to live a moral life, we must know the existence and capacities of the human soul, we must know the reason the soul exists, the role of the body in carrying out the work of the soul, and the destination for which every human soul is made: God Himself. None of these things can be discovered by the physical sciences. Only the theological sciences have the capacity to address these problems.

Theology is the study of these issues, religion is how we live them out. Just as math informs physics, helping physicists know where to look and what to look for, so theology informs religion, helping us know where to look for truth and how to live it.

Living the moral life requires that we first have a clear knowledge of human nature, of divine nature, and of the relationship between the two. Once we know these things, we can be religious, we can live in accord with these truths. By living religiously, we live morally.

8. Has science made belief in God obsolete?

No, science has not made belief in God obsolete. It has, however, made belief in God, and every other kind of person, less immediate.

Experimental science produces increased knowledge of the natural world, which is a most excellent thing. A natural consequence of increasing our knowledge of the natural world is the recognition that the world can be manipulated. We can mold certain aspects of the world to our liking. This is also a good thing.

However, in the course of manipulating the world, we need to create tools, techniques, technology. There is nothing wrong with technology, but we can easily become so fixated on the things we manipulate that we don't end up paying attention to the people sitting next to us.

In a 21st-century computerized world, who has not known a friend, relative or co-worker who gets "lost" in a synthetic world, a wargame, a simulated life? Prior to this, we all knew of people who spent all their time tinkering on car engines or learning new gardening techniques. In every age, it has been possible to get lost in making things do things. Technology is good, but working exclusively with things instead of spending time with persons is not good.

Our current level of technology is very seductive and promises to become more so as time goes on. Working with people is painful – they disagree, they are often intransigent, they refuse to be manipulated like wooden blocks or objects on a screen. All of us are lazy. If we can avoid working with people, many of us generally do.

This distancing of ourselves from everyone else is the beginning of hell. Those in hell believe in God, they just do their best to ignore Him. Those of us intoxicated with science and the things science can produce often just ignore all the persons around us, including God.

We still need personal interaction. But, instead of seeking it out, we simply allow science to prevent us from getting what we need. So, God becomes "obsolete," as do our family and friends, because we convince ourselves that we can get on without them just fine. We can't, of course.

But the illusion that we can, that the next iteration of our technology will finally do the trick, this is an illusion we very much wish were true. In this way, we write ourselves out of existence, we allow ourselves to become submerged in the machine. When we are part of the machine, we don't need God. No machine does. After all, machines aren't persons. And by that point, we are barely persons either.

Science cannot make God obsolete, but it can make us obsolete.

Section 4: Technology

1. Why does the Catholic church have problem with in vitro fertilization?

The Catholic Church does not condone the use of in vitro fertilization because it violates the rights of the embryonic child and it violates the duties of the parents towards their own children. In vitro fertilization is the process by which a human ovum is extracted from a woman's ovary and fertilized with the man's semen. The fertilized egg must grow for several days outside of the woman's body, becoming an embryo, before it can be safely implanted in the woman's womb. Fertilized eggs are incapable of implanting in the womb, only embryonic children can accomplish this task.

Because implantation often fails and thus kills the child, IVF technicians create several children, implanting many, and freezing the rest in a nitrogen freezer.

The embryos are frozen because they are easier to thaw than unfertilized ova are. To date, there is no known way to de-thaw frozen human ova. Thawing frozen human children, on the other hand, kills the child 25-50% of the time. If the embryonic child survives the freezing and de-thawing process, the successful rate of implantation ranges from 0% to 50%, with most success rates being in the high teens to low 20s. In short, for each child who manages to survive the technological gauntlet, the IVF process kills several of his brothers and sisters.

A child has a right to be embraced by his mother from the first moment of his existence. He has a right to be conceived by the loving act of his parents. It is an abuse to allow a child to come into existence outside of his mother's body. It is an abuse to place a child at the enormous risk of death which nitrogen freezing and thawing, and which the IVF procedure itself, constitutes. The IVF process is evil not only because it is so dangerous to children, but also because it turns parents into purchasers, children into commodities and makes the medical community into panderers or slavers, dealing in the commercial creation and sale of human beings for someone else's satisfaction.

> I would say that life has a very long history, but each of us has a unique beginning, the moment of conception... at the very beginning of life, the genetic information and the molecular structure of the egg, the spirit and the matter, the soul and the body must be that tightly intricated because it's a beginning of the new marvel that we call a human.
>
> [L]ove is the contrary of chilly. Love is warmth, and life needs good temperature. So I would consider that the best we can do for early human beings is to have them in their normal shelter, not in the fridge.
>
> ### Jérôme Lejeune
> *Geneticist, pediatrician and devout Catholic.*
> *He was the first to connect a disease, Down's syndrome, to a chromosomal defect, thus opening the whole field of clinical cytogenetics. He has been named "Servant of God" and his cause for sainthood is being postulated.*

2. Why does the Church have a problem with research or treatments involving stem cells?

The Church does not have a problem with stem cells. Indeed, the Church heartily recommends research into the use of adult stem cells for treatment of all manner of human illness and disease.

It is only the use of embryonic stem cells that creates moral difficulties. As seen above, the IVF process produces an enormous number of embryonic children. Embryonic stem cells are obtained by first creating, then tearing apart embryonic children. Creating embryonic children in order to obtain their stem cells for research, or using existing embryonic children in order to accomplish the same goal, is not substantially different from birthing and raising children with the sole or primary view towards selling those children like cattle to a nearby cannibal tribe for their consumption.

It is worthwhile to note that embryonic stem cells have never been used to successfully treat any disease. These cells demonstrate uncontrolled growth, that is, cancerous growth within the human body.

The only stem cells which have proved useful in a clinical or a research setting is adult stem cells. Adult stem cells can be obtained without any danger to the adult involved. They have been isolated from skin, nose and fat tissue, to name but a few locations. When a person's own stem cells are used to treat a disease – a feat which can only be accomplished with adult stem cells, not embryonic stem cells - these cells exhibit no donor vs. graft rejection problems and they don't exhibit uncontrolled growth. Adult stem cells have already been used to treat literally dozens of illnesses.

3. Why is the Church opposed to assisted suicide?

People do not normally seek out death. Death may seem appealing to someone primarily because that person is going through immense suffering, whether physical, emotional, spiritual. Often, this suffering is difficult because it seems useless or punishing. Those who suffer often ask, "Why me? Why not someone else? What have I done to deserve this?" or worse, some may think, "I have done much to deserve this, and would rather not think about it anymore. Release me from my existence so that I am no longer burdened with these thoughts and memories."

It is important to recall that every human person is immortal. Death is the separation of soul and body. Death generally results in the decay and destruction of the body, but the soul survives. Our knowledge, our thoughts, our memories do not die when soul and body separate. Indeed, these are the only things we can carry through the doorway of death, a doorway so narrow that nothing else, not even our own bodies, will fit through.

Thus, suicide is not a solution to the problem of suffering. In fact, it will almost certainly make the problem much, much worse. We don't control the fact of our existence. God does. He holds us in existence from moment to moment. He will hold us in existence for all eternity.

He loved us into existence. He always does what is best for us. Existence is better than non-existence, so He will not allow us to fall out of existence. But, even though God is omnipotent, He will not insult the free will He gave us. If we choose not to be healed of our suffering, He will not impose healing. He will respect our choice. We have to heal the suffering while we are alive, so that we do not carry this suffering with us through that narrow door.

Assisted suicide is generally used by people who don't understand the point of suffering. Often, it is not even really desired by the one being killed. Instead, the ones who watch the suffering decide to put the suffering one out of their misery. The observers are made uncomfortable by this glimpse of their own approaching suffering and death. The scene stirs in them a feeling that they are not ready, not prepared to undergo what the one suffering is now undergoing. Sweep both the suffering and the sufferer under the ground with an injection, a gas mask, a dose of pills, and the discomfort in the observers will leave.

> Pasteur spent years working on a rabies treatment, but no one allowed him to test it. Just as he was about to experiment on himself, a nine-year-old boy, Joseph Meister, was bitten by a rabid dog. The boy's mother begged him to save her son's life. Pasteur injected the boy for 10 days - and the boy lived. Decades later, of all things Pasteur could have etched on his tombstone, he asked for 3 words, "Joseph Meister lived."
>
> ### Louis Pasteur
>
> *Chemist, biologist and devout Catholic.*
> *Demonstrated that life does not spontaneously generate. His work on fermentation, disease and vaccination makes him one of the three founders of microbiology. He died holding a rosary.*

When the sufferer cries out in his agony for an answer, the unprepared observer at first recoils from the cry, then rushes forward to smother the question with a pillow. The one who suffers does not really desire death – he or she desires meaning: "Tell me why I bear this load! Tell me how to bear this load! Help me bear it!"

We must give answers, not silence.

Suffering is a natural evil. It is a lack of something good, a lack that should never have happened to begin with. It is ours because we are a fallen people. But it is not only a lack. Suffering is also hard work, hard work that every person must do at some point.

We can call suffering work because work accomplishes something. Suffering is not a good thing – it is a natural evil. But if we work at joining our suffering to the work of the Cross, our suffering can work miracles in the world. Just as the labors of a parent far into the night can bring joy to a child in the morning, so the labor of suffering now can – if joined to the Cross – help another bear their own load of suffering.

It isn't a question of whether we deserve the particular suffering we experience. It is quite possible that, like Job, we do not deserve it, at least not to the degree that we experience it. But because suffering is work, and we are all one family, God will sometimes allow the stronger among us to bear more suffering precisely because the weaker among us could not stand the strain.

Like Simon of Cyrene, we are sometimes asked to bear a burden that is not properly ours. But if we bear it with the knowledge that this work really makes a difference in the lives of others, really can help others, it is much easier to bear.

4. If Christ saved us by suffering, and we are supposed to imitate Him, then why can Christians use pain-relieving medications?

Many times, the leaders of the people would have stoned Jesus, but He refused to allow it, walking between them and escaping them. He did not allow His suffering and death to occur until the time was right. Similarly, although nearly all of the apostles were eventually martyred, it is also true that when the apostles were persecuted, they fled to another city to preach the Gospel in a new place. They did not accept martyrdom until the time was right.

It is given to men to die once, and then the judgement. If there is good reason to avoid the suffering, if I fear that I am not strong enough to handle the suffering that would otherwise come, there is nothing wrong with fleeing that suffering by using medications which relieve the pain.

God gave us our intellects so that we might use them to bring relief to a suffering world, He gave us to the world to make it a better place. We can relieve suffering in three ways: we can relieve physical suffering, emotional suffering or spiritual suffering.

Physical suffering is the easiest to heal, since it simply requires the use of chemicals. Emotional suffering is more difficult. It requires someone willing to listen, willing to empathize and willing to guide the sufferer out of the emotional morass he finds himself in.

Spiritual suffering is the most difficult suffering to experience and the most difficult to heal. It can be healed only by an infusion of grace. We are not the source of grace, although we can be conduits of grace. In this situation, the best healing comes through the sacraments, which are a direct encounter with God. Sacramentals, like prayer and fasting for someone else, are also of great benefit, although nothing compares with the sacraments.

If we can use the divine sacraments to heal the worst form of suffering – and Christ instituted the sacraments for precisely this reason – then we can certainly use physical analgesics to heal physical suffering when necessary.

5. If in vitro fertilization, especially through the use of donated eggs and/or sperm, is not allowed, then why is organ transplantation allowed?

In organ donation, we are helping to heal a person who is in need of it. With in vitro fertilization, we are bringing a person into existence. Healing an injured person is a substantially different act than helping to create a person.

Now, let's take one cell of a chimpanzee embryo, of a human embryo, of a gorilla embryo and give it to one of my students in the Certificate of Cytogenetics in Paris, and if he cannot tell you this one is a human being, this one is a chimpanzee being, this one is a gorilla being, he would fail his exam; it's as simple as that.

Jérôme Lejeune

Geneticist, pediatrician and devout Catholic.
He was the first to connect a disease, Down's syndrome, to a chromosomal defect, thus opening the whole field of clinical cytogenetics. He has been named "Servant of God" and his cause for sainthood is being postulated.

When we bring a person into existence outside of the womb, we are treating that person with disrespect. We have a right to originate within the act of interpenetrating love between a man, a woman and God, not through the rote actions of a bored lab tech with a pipette and a Petri dish.

Organ donation can be a great gift, but it can only be made under circumstances that guard the respect due to everyone involved in the donation. For instance, the donor cannot morally make a donation if that donation would kill the donor. It is only possible to be a living tissue or organ donor when the tissue or organ in question can be safely donated. Thus, since blood regenerates relatively quickly, as long as only limited quantities are taken it may be donated. The loss of one kidney does not substantially impair the life of the donor. The donation of one lobe of the liver would be acceptable, etc. But a living donor could not donate his all of his blood, both of his kidneys, or his entire liver. Organ transplantation in these scenarios would be morally impossible. We would be treating the donor as an organ warehouse, not respecting him or her as a person.

Similarly, it is not moral to kill someone else, even if that someone else is sick or dying, in order to harvest their organs for transplant to others whom we deem more worthy of life. It would be profoundly disrespectful to everyone involved if we did.

People are not tools to be used, they are not to be bought and sold, or brought into existence like any other commodity. They are persons to be honored. The beginning and the end of life are times where it is particularly easy to run roughshod over the frailty of the individual. It is very easy to say, "There is no person here. In any case, in a little while, there won't be."

The refusal to recognize the presence of the person is a refusal to recognize the sign of God's presence. He is the original three Persons – we are made like Him so as to remind each other of Him. If we fail to do this, we are turning our backs on God.

6. If artificial hearts, lungs and limbs are allowed, could robots ever be considered persons? How much can be replaced before there is no more "person" there?

Two organs cannot be transplanted: the human brain and the human gonads. The human brain cannot be transplanted because it is the tool through which the human soul expresses itself in the world. It is unique to each person, allowing a unique expression of each person.

Similarly, the gonads are the source of the body, the tool by which the body makes a gift of itself to the next generation. It uniquely carries the individual's bodily identity in its DNA. Transplanting my gonads into your body, or yours into mine, would mean that the one receiving the transplant would be lying with his body. When I had sex, I would be pretending to make a gift of my own body to the child being conceived when I am, in reality, actually transmitting your body instead. We would be pretending to transmit our genes to our children in the sexual act, when we were really not transmitting our own genes at all.

Thus, while heart, lungs, kidneys, muscle, skin, cornea, bone, and sundry other body parts can be morally transplanted or replaced with artificial substitutes, the brain and the gonads could not be. Since we cannot create souls, we have no way of making a robot a person.

7. What end of life treatment must we accept?

Life is a very great good, but it is not the ultimate good. If we are pursuing a greater good, we are permitted to relinquish the lesser good in order to attain the greater. The greatest good is attaining union with God. So, if I know people who are not headed towards God, I am permitted to give up everything, even my life, in order to help those people understand Who God Is and how much God loves them. This is why martyrdom is permitted – I am allowed to give up my life in order to bring about the greater good of helping someone else towards salvation.

Suffering and death are both natural evils. We must pass through both. Neither can be avoided. So, even though life is a very great good, we are permitted to take whatever steps are necessary in order to die in a state of grace and thereby assure salvation. That means we must die in such a way that we harm no one, that we take no innocent life, not even our own. We are not required to take extraordinary measures to stay alive, but we are required to use ordinary measures for as long as they work.

Since no person can survive without food and water, since cutting off food and water to anyone assures their death, we may not refuse food and water under normal circumstances. Cutting off food and water would be murder if I did it to someone else, suicide if I did it to myself. The only reason we might have for being able to refuse either is if our bodies are so far along the path towards death that they are incapable of actually using either.

So, if we can no longer swallow, it is now considered standard care to insert a feeding tube. The only reason a feeding tube could be refused is if the food is not being broken down by the body, it isn't being used by the body and it is making the patient uncomfortable. Similarly, the only reason an IV drip to hydrate the patient could be refused is if the kidneys had shut down and the additional fluid was simply causing the patient to blow up like a balloon – the fluid wasn't being utilized and it was making the patient uncomfortable.

But if food and water are being used, then both must be supplied. A heart or lung machine, on the other hand, can be refused. If the patient is not able to breathe or pump blood on his own, then he is very near death. Removing the machine is not murder in this case because the patient's body has begun to shut down, it is very close to death.

We should accept whatever treatments are necessary for us to put our affairs in order, speak to our family, receive the sacraments of the Church. Now, at times, certain pain-killers can hasten death. Although we know this is the case, the use of these pain-killers are permitted if the resulting state of lowered pain allows the patient to wrap up his affairs, say his goodbyes and receive the sacraments. The pain-killers cannot be given in order to cause death, rather, they must be given in order to allow the patient to live what is left of his life so as to accomplish the work that is before him.

8. Can God draw a square circle? Can God make a rock so big He can't lift it?

God is reality, He isn't fantasy or illusion. God is completely rational. However, He is also three Persons. Because He is three Persons, He is more than just logic, yet everything He does is logical.

Now, the definition of a circle is "the set of points that are equidistant from a given point." This definition cannot be made logically identical to the definition of a square, a parallelogram with equal sides and one right angle. Since God is too great to contradict Himself, He does not do so.

A square circle is a contradiction in terms. Just because we can string words together into a sentence doesn't mean the sentence makes any sense. God doesn't do nonsense.

Isaac Asimov, a scientist and a devout atheist, pointed out that the second question is equally self-contradictory. He phrased it in a slightly different way, a way his beginning physics students often used to needle him, "What happens when an irresistible force meets an immovable object?"

Ironically, Asimov didn't realize he was not just answering a physics problem, he was also answering a theological problem. This is his answer:

As Einstein pointed out, mass is energy.

Each is just a different form of the other, something like water and ice. Matter is frozen energy, energy is liquid matter. So, can an "irresistible force" meet an "immovable object"?

No.

The force can only be irresistible if it comprises the majority of the energy in the system. But since energy and matter are interchangeable, an object can only be immovable if it, too, comprises the majority of the energy in the system.

There can't be two majorities in a system.

Thus, as Asimov points out, the question is really nonsensical. It is another example of stringing words together without actually saying anything.

Section 5: History and Philosophy

1. Why did the Church persecute Copernicus?

She didn't persecute Copernicus. Rather, the Church provided financial support to Copernicus for his whole life, and honored his astronomical expertise when it came time to reform the calendar. It was the academics who threatened to persecute Copernicus.

Copernicus was a canon of Frombork Cathedral, trained in Catholic universities with a doctorate in canon law and a degree in medicine. His maternal uncle, Lucas Watzenrode, Bishop of Warmia, supported Copernicus' astronomical work financially for Copernicus' entire life.

> My goal is to find the truth in God's majestic creation.
>
> **Nicolas Copernicus**
>
> *Astronomer, mathematician, canon lawyer and Catholic priest. He developed the heliocentric theory and assisted with the Gregorian reform of the calendar.*

For Father Copernicus, church business, especially economics and diplomacy, was his life; astronomy was his hobby. But his astronomical skill was well-known. In 1514, the Fifth Lateran Council sought his opinion in the problem of calendar reform, giving him occasion to publish his first heliocentric theories. Indeed, the Gregorian calendar reform of 1586 was based on heliocentric theory. These theories were so elegant that in 1533 they were presented in Rome to Pope Clement VII and several cardinals. On 1 November 1536, Archbishop of Capua Nicholas Schönberg wrote Father Copernicus a letter in which he strongly encouraged him to continue to develop these new cosmological theories and offered to help pay printing costs. But, though Bishop Giese also urged him forward, Copernicus held off on publication. He was uncertain how the secular university professors, who held fast to Ptolemaic astronomy, would react to his ideas. His fears were not unfounded. When his work was presented at the University of Wittenburg, the faculty professors there were so hostile to his work that they permitted only the chapter on trigonometry to be printed.

But, in stark contrast to the academic reception, Bishop Giese was so interested in having the work published that he employed George Rheticus, a Protestant pupil of Copernicus and a man whose father had been beheaded for sorcery, to edit Copernicus' manuscript. The secular professors of Wittenberg were incensed. They silenced Rheticus by forcing him to give up his chair in mathematics. As a result, Rheticus was forced to give up the editing project, which he handed over to Reverend Osiander, also a Protestant.

Osiander, aware that Luther and Melancthon hated Copernicanism even more than the secular university professors did, tried to convince Copernicus to write a preface which called the solar-centered universe a mere hypothesis. Copernicus not only refused, he instead wrote a wonderful preface dedicating the work to Pope Paul III. Since Copernicus was very ill he relied on Osiander to get the manuscript ready for printing. Osiander secretly replaced Copernicus' preface with his own unsigned preface, falsely claiming the contents of Copernicus' book were hypothetical and not to be taken seriously. Copernicus, who only saw a bound copy of the work as he lay dying from the effects of a stroke, didn't realize what had been done.

Other astronomers knew. Johannes Kepler, for instance, took the time to demonstrate that the preface was a forgery created by Osiander out of fear of Protestant and secular academic reaction. The war against Copernicanism and heliocentrism was begun by university professors, not the Catholic Church.

So, the only resistance Copernicus encountered during his life was from Protestants and lay university professors. It is highly ironic that university professors today invoke Copernicus as one of their martyrs, given that he wasn't martyred, and what difficulties he did experience came from the university.

2. Why did the Church persecute Giordano Bruno?

To understand what happened to Bruno, we must first know who and what he was. Giordano Bruno was ordained in 1572 as a Dominican priest. He studied humanity, logic and dialectic, not astronomy. While he enjoyed mathematics, he does not seem to have engaged in any scientific observation or work during his lifetime.

Even as a novice, he was well-known for expressing strange theological views. By 1576, he had been formally accused of heresy. He went to Rome to teach at a convent, but the nuns liked his teachings no better and he was accused of heresy again. He then left the Dominican order and began wandering from town to town, angering every person who offered to help him.

Although he later denied that he had ever been a Calvinist, the Calvinist Council of Geneva excommunicated him. Leaving Geneva, he went to Toulouse and Paris. Here he published a memory-training system he had developed, along with a scatological work intended to be humorous. From Paris, he moved to England, where he was hosted by the Protestant Queen Elizabeth, refused a professorship at Oxford, and wrote a scathing attack on the Oxford dons. Returning to Paris, he tried to reconcile with the Church but couldn't manage it because the Church required that he return to his order, which he didn't want to do.

By 1587, he was in Germany, where the Lutherans excommunicated him. He was invited to Venice in 1591, on condition that he teach his host his memory-training system. Unfortunately, he went to Venice but then refused to train anyone. His host angrily denounced him to the Inquisition. There he attempted to defend his understanding of Catholic doctrine and practice. In 1593, he was extradited to Rome, and held until 1599. Although given several opportunities to retract his errors, he was finally tried and condemned as a heretic in January, 1600 and given to secular authorities. The state chose to burn him at the stake on February 17, 1600.

The trial summary shows that he was not condemned for holding to heliocentrism nor for his theory that there were an infinite number of worlds (although he *was* asked to abjure this assertion), but rather because of his theological errors. The condemned propositions which he held included: Christ was not God, but only a really skillful magician, the Holy Spirit was the soul of the whole world, the devil could be saved, human souls could transmigrate into the bodies of animals, God and the world are one, everything in the universe is alive, including rocks, mountains, etc. He thought matter and spirit, body and soul were essentially the same substance. He believed in magic.

The laws of nature are written by the hand of God in the language of mathematics.

Galileo Galilee

Physicist, mathematician, astronomer and devout Catholic. He made major contributions in hydrology, dynamics, astronomy and mechanics. Many consider him the Father of Modern Science.

Proof that Bruno was not on trial for heliocentrism can be found in the fact that, though Bruno was burned in 1600, priests, bishops and cardinals did not hesitate to throw Galileo parties to celebrate his telescopic observations in 1611. Further, Bruno's name never come up during either of the Galileo trials. Since Cardinal Bellarmine oversaw both the trial of Giordano Bruno and the first trial concerning heliocentrism in 1614, this is rather a significant omission. If Bruno's heresy had involved heliocentrism, neither Bellarmine nor the Pope would have permitted the parties in Galileo's honor in 1611, while during the investigation of 1615, Bellarmine would certainly have told the parties involved that heliocentricism had already been condemned by himself during Bruno's trial just 14 years before. He did neither of these things.

In fact, Bruno never actually performed any scientific observations or measurements, never formally studied the sciences, and although everyone agreed he was very intelligent, he was not recognized by any of his contemporaries as a scientist. His work can be summed up as a series of fairly wild and definitely unsubstantiated philosophical tracts.

Giordano Bruno is not a martyr for science.

3. Why did the Church persecute Galileo?

The Church was goaded into persecuting Galileo by university professors. Contrary to popular belief, neither Copernicus nor Galileo were initially attacked by the Catholic Church. Indeed, both received most of their initial support from Catholic priests, bishops and popes. When it came to these two mathematicians, it was the lay academic community, the university professors, who hated their guts.

Galileo, you see, had the unfortunate distinction of being a mathematician at a time when mathematicians were universally considered second-class citizens by the academic community. Mathematicians were shady characters, good only for creating siege engines, building fortifications and casting horoscopes. Galileo was so well loved by his colleagues that he was run out of the University of Pisa, and as the chair of mathematics at the University of Padua, he earned less than one-tenth what the best-paid Aristotelian philosopher earned.

Aristotelian philosophers were at the top of the lay academic pecking order primarily because Aristotle's physical theories were based not on mathematics but on philosophy. He assumed that every inanimate object had an innate purpose that determined its motion. Rocks fell down because they intended to reach the center of the earth. Hot air rose because it intended to reach the celestial sphere. Since intention determined the direction of motion, and since philosophy is the key to understanding purpose and intention, philosophy was considered the best way to understand the workings of the universe.

A mere mathematician could never hope to plumb the universe's depths of mystery. Mathematicians played with mindless numbers that had little to do with the real world. They cast horoscopes for superstitious people, which was a mortal sin. They often dabbled in the occult. Their number-play was only good for creating accurate calendars and calculating where a cannonball might land.

Heliocentrism was opposed by the academics not because it was a violation of Scripture so much as it was a violation of Aristotle. Any theory that seemed to contradict Aristotle also contradicted the authority of the university professors. In short, a sun-centered universe directly attacked the prestige of most of the lay academic community.

> You have been disarming by steps those who have control of the sciences (the university academics) and they have nothing left but to run back to holy ground.
>
> **Francesco Niccolini**
>
> *Ambassador, Grand Duke of Tuscany, and political advisor writing to the Duke of Florence's court mathematician, Galileo Galilee.*

Consequently, the most vociferous opponents to heliocentrism would be the members of the academic community. Copernicus was fortunate enough to die the same day his book was released from the printer. As a result, he did not have to face the abject hatred poured out on his work by the university community. Galileo, however, saw the vitriol poured out by academia upon the dead Copernicus and hated them for it.

He ridiculed his fellow academics from the very first moment he began lecturing at Pisa, writing poetry that made the academic gowns the laughing-stock of the town. His short tenure in the mathematics chair at Padua was not much more successful. Few people today remember that Galileo did not work for a university when he made his most important discoveries. The academics had driven him out long before. Instead, he found private employment with the Count of Florence. Galileo hated the university professors as much as they hated him.

Thus, when Galileo's telescope brought supporting evidence for the Copernican theory, it was not the Church that attacked him, it was the academic community. Even as the Jesuits and Dominicans threw luxuriant parties for Galileo in Rome to celebrate his new discoveries, the lay academics schemed to destroy this disrespectful upstart, this dirty little *mathematician*. Indeed, while priests and bishops delighted in the new vistas the telescope opened up, most of the academic community refused even to look through the lens. They claimed the visions within were nothing but optical illusions. Magini, the university professor and famous Ptolemaic astronomer, promised to wipe Galileo's new planets from the sky.

Unfortunately, Galileo's fight unfolded less than 100 years after Luther had nailed his 95 theses to a Wittenburg church door. As Protestants vied with the priests and bishops over the proper interpretation of Scripture, the academics saw their chance to take down their hated enemy. It was the lay academics who first brought Scripture into the heliocentrism debate, accusing Galileo of heresy, of violating the God's own divine word.

It was the lay academics who duped a foolish Dominican priest into attacking Galileo from the pulpit, much to the dismay of the Dominican astronomers who had just celebrated the astronomer from Florence. The Church was eventually drawn into the controversy by university professors, advisors to the Church, men who insisted that Rome had a duty to stop Galileo, for he were left unchecked, he would destroy the entire university system.

They were half-right. While Galileo's theory was not dangerous to theology, Galileo did, indeed, destroy the Aristotelian philosophy professors. He became the first man in history to use mathematics to systematically and completely describe the way the objects in the world interacted with one another. He stripped away the false Aristotelian idea that we must first understand an object's purpose before we can understand how objects interact. He showed that mathematical formulas alone were sufficient to describe movement. In short, he proved that inanimate objects were truly inanimate — they were not quasi-persons with intentions or purposes. Galileo drove the last nail into the coffin of Aristotelian paganism. He also drove a stake through the heart of every philosophy department in every university in Europe. After Galileo, philosophers would never again regain their former stature, nor would they ever again be able to claim that philosophers alone were able to describe and explain the universe.

4. Is the persecution of Galileo unique?

Galileo's experience is essentially unique in the life of the Catholic Church. No other scientist was ever called before the Inquisition or had his scientific views investigated as he did. However, the persuection of scientists is certainly not unique. It is generally a privilege scientists reserve to themselves. They dislike it when others tread in their domain.

Consider, for instance, the case of Ignaz Semmelweis, the man who discovered that a doctor who has just done an autopsy should wash his hands before he does a gynecological exam of a pregnant woman. Through careful documentation of Viennese maternity ward deaths, Semmelweis demonstrated that failure to wash hands between doing autopsies and examining maternity patients directly correlated to the number of pregnant women who died in the maternity wards. He insisted it was the doctors who were causing the deaths.

As with Galileo, Semmelweis' colleagues refused to believe it. Although Semmelweis had used his hand sterilization techniques to cut death in his maternity wards by over 50%, his theory predated germ theory. His explanation for why hand-washing reduced death - it removed some unknown disease-causing essences from the hands - was considered ridiculous. The other doctors on the hospital staff invented all kinds of reasons to deny Semmelweis' findings, going so far as to blame the mortality rate on the fact that a priest often walked through maternity wards. The fearsome sight of the priest was apparently enough to induce childbed fever and subsequent death.

Not only did they refuse to accept his theories, the medical profession persecuted him so vigourously that he was reduced to poverty and eventually committed to an insane asylum. Within days, he was dead from the beatings.

Georg Cantor's work on the problem of infinite sets in mathematics was treated with similar respect. Cantor invented set theory to handle the problem of transfinite numbers. Himself a devout Lutheran, he considered set theory to be direct proof for the existence of God.

I am not been the inventor of the ideas about infinity that I have published. I am merely a mouthpiece, inspired by God to communicate parts of the mind of God to everyone else.

Georg Cantor

Mathematician and devout Lutheran.
Created set theory, a foundational branch of modern mathematics.

Other scientists were not so kind. His work was attacked as "pernicious," "a grave disease," and "utter nonsense" while he himself was called a "scientific charlatan," "renegade," and a "corrupter of youth." He was driven into depression and eventual suicide. Ludwig Boltzman, the man who essentially created the fields of statistical mechanics and statistical themodynamics, suffered a similar fate.

But scientists are not alone in successfully persecuting their fellows. Politicans also get in on the act. After World War II, the US government tried J. Robert Oppenheimer, the father of the atomic bomb, for holding views that appeared to be in discord with the contemporary political climate, much as Giardono Bruno had been tried. During the French Revolution, Antoine Lavosier, considered the father of modern chemistry, was executed by the French revolutionaries who asserted, "The Revolution has no need of scientists." Similar executions were carried out by Josef

Stalin upon biologists who refused to accept the biological theories of Trofim Lysenko. Despite the similarity of these events, Copernicus, Bruno and Galileo's names have become by-words of religious oppression. Yet no one mentions the political oppression suffered by Lavosier and Oppenheimer, while the martyrs killed for opposing Lysenko are so numerous they have become essentially anonymous.

But it doesn't stop there. Scientists themselves have used science to persecute and kill countless men, women and children. While many refuse to admit it, the German scientists did real science in their death camp experiments during World War II. It is a fact that Nazi hypothermia experiments, in which the Nazi scientists deliberately killed people by immersing them in ice water or exposing them naked to bitter cold in order to study the effects, produced data so valuable that the Allies used that same data to treat and combat hypothermia in the years following the war. Similar use was made of the Nazi data on phosgene, a poison gas, and the data from biological warfare experiments done on American prisoners by the Japanese during the war.

And, lest we think that only scientists employed by dictators would do this kind of thing, let us consider America's own little shop of scientific horrors. The most famous is undoubtedly the Tuskegee syphilis experiments. From 1932 to 1972, medical doctors deliberately allowed poor black men to die of untreated tertiary syphilis so the effects of the disease could be studied.

In 1949, American scientists introduced radioactive material into breakfast cereals and fed them to children at the Fernald School for the mentally retarded in Waltham, MA, so as to study the effects of radiation on these "useless eaters." In Cincinnati, impoverished cancer patients were subjected to whole body radation in a 12-year experiment that started in 1960. All children admitted to Willowbrook State Hospital in New York after 1964 were injected with hepatitis in order to study its effects. In 1963, patients at the Jewish Chronic Disease Hospital were deliberately injected with cancer cells to see how long the foreign cells would survive. In 1999, *all* research projects at the Veterans' Administration West LA Medical Center were shut down because its doctors refused to stop experimenting on patients who refused consent.

> Teach us to study the works of Thy hands that we may subdue the earth to our use and strengthen our reason for Thy service.
>
> **~ *James Clerk Maxwell***
>
> *Theoretical physicist, mathematician and devout Presbyterian. He contributed to the fields of optics, color vision, gases and elasticity but is most well-known for his work in uniting electromagnetic theory with light theory. His work is the basis for all modern physics. He was also a devout Christian deeply familiar with the Scriptures.*

Thomas Huxley, a man so fanatical about the defense of Darwinism that he is to this day known as "Darwin's bulldog," examined the Galileo case and opined, "The Pope and the Cardinals had the best of it." He recognized that the line they took in regards to Galileo had ultimately been fair. The Church insisted Galileo treat his heliocentrism as an opinion until decent proof was provided. This insistence was made only after tremendous provocation had been offered by the academics who opposed Galileo. Galileo was not put to death, he was not tortured; his sentence was a requirement to pray the penitenital Psalms (his daughter, a cloistered nun, was permitted to do this penance for him), and a "house arrest" so lenient that Galileo was able to publish his greatest work, on mechanics, while living under its burden. Indeed, it is this last work, not his book on heliocentrism, which makes him the father of science.

Compare his fate to that of any of the others related on these pages. While the Catholic Church has officially apologized to the world for the aspects of Galileo's case which were unjust, we have not heard a word of apology from anyone else.

5. What is the relationship between science and philosophy today?

Galileo destroyed the chairs of philosophy. They have never regained the highest place of honor in the pantheon of human knowledge. But, since Galileo's time, the rising scientific community has made its own egregious error. As it gained ascendancy and public adulation, it has continued to attack and abjure the necessity of philosophy and theology. Galileo gained fame by driving purpose out of the study of rocks. Many scientists today believe they can drive purpose out of the study of persons.

Unfortunately for promoters of experimental science, philosophy cannot be completely eradicated. The study of purpose - the study of philosophy and theology - is unavoidable. The mathematical method of studying the world itself embodies a philosophy, and a remarkably incomplete philosophy at that. Numbers can only tell us what, they can never tell us why. Numbers describe but they do not ultimately explain.

Science is about nothing but numbers — measurement is the foundation of everything it does. Because it focuses so doggedly on numbers, many of its proponents have begun to insist that there is nothing beyond numbers — there is no purpose, no intentionality, nothing beyond measurement and description. This is the philosophy of atheistic scientists in a nutshell.

Any theory which attempts to provide a volitional explanation is derided by them as mere philosophy, or worse, religion. Thus, today, the same battle lines are being drawn as were drawn against the staunch Catholic scientist, Galileo. Once again, we see the academic community arrayed against the philosophers and theologians. This time, however, the roles are reversed. Now the scientists possess the heights of adulation, while the philosophers and theologians are paid a pittance in both salary and respect.

In Galileo's time, the philosophers hung grimly onto their posts by denying the use of mathematics and insisting that only purpose mattered. Today's scientists hang grimly onto their posts by denying the importance of purpose and insisting that only measurement matters.

Today, both sides fail to realize the essential complementarity of science and theology. Science describes the relationship between objects. Theology describes the relationship between persons. Because persons possess bodies, that is, because persons can be treated as objects, science makes the fatal mistake of assuming persons are objects. Because science is so successful at describing the interaction between inanimate objects, scientists they think they can successfully describe the interaction between persons.

Personal relationships are not subject to what scientists do best. Scientists measure.

So, how much do you love your wife?

14294.232?

π?

Numbers cannot be assigned to relationships. Quantity is most certainly a quality, but quantity does not exhaust every quality a person may reveal. The qualities of inanimate objects can be revealed through external study, but the qualities of a person are revealed only through self-revelation. We can see what a person does, but we cannot know with certainty why the person does it unless that person reveals the why. What we cannot ask of objects — the why — we cannot refrain from asking of subjects, of each other.

Objects are not known through self-revelation. But persons are known only through self-revelation. The inquiry into the origins of persons cannot be solved through external study alone, because the very definition of person assumes a presence that is beyond the reach of even the most delicate scientific measuring instruments. These points are too often lost on everyone in the debate.

Thus, just as the university professors of Galileo's time used Scripture as a weapon to attack the scientist, so today's scientists use Scripture to attack the philosopher and theologian, but in an oddly perverse way. The original attack was built on the immutable authority of Scripture. Today's attack is built on the supposition that Scripture has no real authority, and anyone who adheres to it is, in fact, a fool and an ignoramus of the first order.

In modern times, Scripture lacks authority in part because Scripture does not measure. It is not experimental. It is not "scientific" – forgetting for the moment that there is more to science than is dreamt of in a Bunsen burner.

natural phenomena, or the prophecies weren't prophecies at all. According to this view of things, the Jews were really just writing down events after the fact, pretending that these events had been prophesied by their forefathers.

The grounds for this assertion seems to be rooted not just in disbelief, but in something a little more unsavory. In fact, they really appear to be nothing more than a covert anti-Semitism. After all, how are such theories different from saying, "Well, everyone knows the Jewish tendency towards deception, either of themselves or of others."?

When we consider the question of prophecy more carefully, we can see prophecy is not a very distant cousin from something experimental scientists do every day - they predict events. These predictions are based on factual observation. Which way will the hurricane travel? How will the mouse respond? Will this antibiotic work against this bacillus?

When, on 5 June, 1876, Le Verrier presented to the Academy his completed tables for the movement of Jupiter, the result of thirty-five years of toil, he emphasized particularly the fact that only the thought of the great Creator of the universe had kept him from flagging, and had maintained his enthusiasm for his task.

Urbain-Jean-Joseph Le Verrier

Mathematician, Astronomer and devout Catholic.
The first man credited with the discovery of a planet, Neptune,
through mathematical calculation alone.

Scientists discover the answers by running an experiment with a control: they set up two identical situations, change one thing in one of them, and then observe the results. If the results for each situation are identical, then the change was unimportant. If the results are different, the difference is somehow due to the change.

Petrus Peregrinus, the 13th century Franciscan friar, formally described the scientific method of induction. Galileo popularized it to the world three centuries later. But even though it was developed in the thirteenth century, it is interesting to note that the method was not original to that period. Rather, the methods of scientific testing grew out of a specific Judeo-Christian worldview.

7. Many religions claim to have actually laid the groundwork for some important finding of hard science. Why should we believe Jews or Christians are the foundation of science itself?

Look at the evidence and decide for yourself. Remember the scientific method: set up two identical situations, change one item, and observe the outcome. If the outcome between the two situations is different, it is probably due to the change.

Read Judges 6:33-40. Decide for yourself if Gideon used essentially this method when he set out a sheep's fleece on the ground two nights running and prayed that God might grant two radically different results. In order to help him discern God's will, he made the two situations as identical as he could and varied only one thing: his prayer.

Read 1 Kings 18:1-46. Decide if Elijah used this same method in his argument with the prophets of Baal. He told the prophets of Baal how to set up a controlled experiment - dual sacrificial altars, one to his deity and one to theirs - in order to demonstrate which God was the real one.

Consider Numbers 16. Decide if Moses demonstrated a similar interest in controlled experiment. When Korah questioned whether Moses was really different from any other man who prayed to God, Moses replied by instructing the 250 men who followed Korah that he and they should all offer identical sacrifices of incense to God. At the end of the test, the differences were clear.

Read the first chapter of Daniel. Decide if similar research into the utility of vegetarian versus meat-rich diets is conducted there. After a ten-day trial diet, in which the only variable was the presence of meat, Daniel proved to the king's man that a vegetarian diet was beneficial.

Consider how Pharoah felt about the importance of reproducibility when he tested Moses' and Aaron's miracles. When Aaron's staff became a snake, when the river turn to blood, and when frogs were called forth, Pharoah was unconcerned. Precisely because his own court magicians could do the same thing, Pharoah had no reason to believe any of these events came from a uniquely powerful god. It was only when Moses called forth lice that the court magicians said, "We can't do this. This is the finger of God."

But, as is so often the case with government-sponsored research, the politician didn't want to hear it (Exodus 7 and 8). Pharoah ignored the conclusions of his own researchers. He continued to refuse the idea that Moses' God was unusually powerful. In dozens of passages, time and time again, the Hebrew Scriptures insist on the importance of the test: God tests His people constantly, and His people test Him. Indeed, it is through this constant testing that His people come to trust Him, proclaiming Him "Always Faithful" (Psalm 89:28)

The test is central to the Hebrew understanding of how they interact with both nature and nature's God. But the Hebrew Scriptures do not simply prove the importance of tests and the scientific method. These Scriptures also make several other assertions crucial to the development of experimental science.

Measure what is measurable, and make measurable what is not so.

Galileo Galilee

Physicist, mathematician, astronomer and devout Catholic. He made major contributions in hydrology, dynamics, astronomy and mechanics. Many consider him the Father of Modern Science.

As the physicist and priest, Dr. Stanley Jaki, liked to point out, it is precisely the idea that "God disposed everything according to measure and number and weight" (Wisdom 11:20) that drives every experimental science. If it cannot be measured, it is not subject to experiment. Furthermore, these same Hebrew Scriptures imply that tests are properly applied to nature because:

> The heavens show forth the glory of God,
> and the firmament declareth the work of his
> hands./Day after day they utter speech, and
> night after night display knowledge. / There
> are no speeches nor languages, where their
> voices are not heard. / Their sound hath gone

forth into all the earth: and their words unto
the ends of the world. (Psalm 19:1-4)

But even with all of this support, the Hebrew people
did not develop a full-bore experimental science. What else
was missing?

The Jews who followed a man named Jesus supplied
the final philosophical piece. In asserting the incredible
claim that God took on human flesh, these Jewish
"Christians" made a claim that was eminently testable.
The men who made this claim knew that full well. As Paul
himself insists, Christians have to make rational assessment
of the allegations he makes, because he fully understands
what is true if he is wrong:

> If there is no resurrection of the dead, then
> Christ is not risen. And if Christ is not risen,
> then our preaching is empty and your faith
> is also empty. Yes, and we are found false
> witnesses of God, because we have testified
> of God that He raised up Christ, whom He
> did not raise up—if in fact the dead do not
> rise. For if the dead do not rise, then Christ is
> not risen. And if Christ is not risen, your faith
> is futile; you are still in your sins! Then also
> those who have fallen asleep in Christ have
> perished. If in this life only we have hope in
> Christ, we are of all men the most pitiable.
> (1 Corinthians 15:13-19)

Testing is central to Paul's theology. His letter exhorts
the Thessalonians: "Test everything, hold fast to what
is good." (5:21) His secretary, Luke, the author of both a
Gospel and of Acts, constantly reminds the reader of all the
witnesses who saw the events he and Paul describe. Luke

even points out he painstakingly interviewed eyewitnesses, thereby testing their testimonies against one another.

John's first letter likewise exhorts Christians "Beloved, do not believe every spirit, but test the spirits to see whether they are from God, because many false prophets have gone out into the world." (1 John 4:1). The author of the Gospels and of Acts constantly point to the number of witnesses still alive who can personally testify to:

> That which was from the beginning, which we have heard, which we have seen with our eyes, which we have looked upon and our hands have handled, of the word of life. For the life was manifested: and we have seen and do bear witness and declare unto you the life eternal, which was with the Father and hath appeared to us. That which we have seen and have heard, we declare unto you: that you also may have fellowship with us and our fellowship may be with the Father and with his Son Jesus Christ. And these things we write to you, that you may rejoice and your joy may be full. (1 John 1:-4)

When the Hebrew emphasis on test, measurement, and natural evidence is combined with the skepticism enjoined upon the Christian, we see the very heart and soul of the scientific attitude. With the invention of the printing press, this flowered into full-blown scientific community.

Did Peregrinus and Galileo have these Scriptural examples, this Scriptural philosophy in mind when they formally described how experiments should be done? We don't know. But we do know that Peregrinus, being a Franciscan friar, knew the Scriptures intimately, prayed them as part of his daily office of prayers. We also know Galileo was a staunch Catholic who was very familiar with Scripture, as were his daughters, one of whom entered a convent.

8. But what of the Scripture that says "faith is the evidence of things not seen" and "Blessed are those who have not seen, but still believe"?

Let's take the last Scripture first, wherein Thomas was apparently rebuked for wanting to put his fingers into Christ's wounds before he would believe in the resurrection. Is this not a repudiation of everything that was just said?

Actually, no. This complete mis-reads the text. Thomas was not rebuked for wanting physical evidence. Indeed, Christ explicitly told him to go ahead and get the physical evidence - "Put your finger here, see my hand. Reach out your hand and put it in my side. Do not disbelieve, but believe" (John 20:27). Thomas was specifically told to trust the evidence of his own senses.

So why was he rebuked? Christ rebuked Thomas not for wanting this physical evidence, but for refusing to believe his friends, the apostles. God is Three Persons. The point of taking on flesh was to teach human persons to learn to trust and love other persons, especially the Divine Persons who love us into existence and care for us every day. God revealed Himself in such a way that we would be forced to learn to trust each other.

After the Ascension, the only way to verify the fact of the Resurrection would be through the testimony of witnesses who had done what Thomas had just done, seen what Thomas had just seen. Indeed, Luke tells us the risen

Christ appeared to hundreds after His death, and in every appearance, He always provided the witnesses with physical evidence of His Resurrection (Acts 10:41). If Thomas refused to believe the witnesses, then others would refuse as well. As with the science of law, theology is about interpersonal relations, about a person's testimony. Theology requires the witness of persons. Thomas had doubted reliable witnesses - that's bad theology. Jesus rebuked Thomas for his lousy theology, not for his experimental science.

As the depth of our insight into the wonderful works of God increases, the stronger are our feelings of awe and veneration in contemplating them and in endeavoring to approach their Author ... So will he [the earnest student] by his studies and successive acquirements be led through nature up to nature's God.

~ William Lord Kelvin

Mathematician, physicist, engineer and Presbyterian. Formulated the first and second laws of thermodynamics and helped lay the first trans-Atlantic cable. He was an elder of St Columba's Parish (Church of Scotland) in Largs for many years before his death. The Kelvin temperature scale is named for him.

And, to be perfectly fair, Thomas' approach to theology is not ultimately good for science either. If science is to advance, it requires very much the same kind of trust in friends and acquaintances that Thomas had failed to show in his fellow apostles. Consider: is theological trust really so different from what the experimental scientist does?

Yes, we know scientists pride themselves on taking nothing for granted. But it isn't true. After all, has a scientist performed *every* experiment upon which his science is based? No, he has not. As a student, he reads and largely accepts what is in his textbooks. As a professor, he reads and largely accepts what is in his journals - peer-reviewed journals, written and vetted by persons who have his trust. He trusts those reporting the experiments have accurately reported the facts. He trusts that the peers reviewing the work have correctly discerned its veracity. He commits acts of faith in his fellows every day or he could learn very little and accomplish even less. Theologians do much the same.

This also explains the Hebrews passage concerning faith and evidence. Blind faith doesn't exist. It is a contradiction in terms. Hebrews does not say, "Faith requires no evidence." Rather, it says faith *is* evidence - it is evidence of something else. Faith is one person's witness that he saw or heard a test, or testimony, that helped him make sense of the world. It counts as evidence. But we already know this. We already agree with this.

The simple act of ordering a hamburger at the local fast food joint is an act of faith. Think about it. When you drive down the road, hungry, you stop at a particular store because it advertises itself as being able to supply food.

Do we know that it will be able to provide? Not really. Perhaps the sign is wrong (new ownership this morning has turned it into a hardware store and they haven't had time to change the sign). Once we get inside, we see a menu, people eating food, plates, forks, napkins, condiments, and someone ready to take our order.

But can they supply food? We don't honestly know that they can. Perhaps the previous customer just got the last hamburger and fries. Perhaps the cook has just gotten sick as we walked in, and is no longer capable of preparing the meal. Perhaps the cook is incompetent and the meal will be inedible when we finally get it.

Yet we give our money to the man behind the counter before we have even received the food, tasted the first bite. We expect to get food. That's faith, and it is based on evidence, as all real faith is.

If we entered a hardware store, saw bins of nuts and bolts, heard the clerk explain to a customer how to install a sink, and then we asked that hardware store employee for a double cheeseburger, fries and a shake, *that* would be an act of blind faith. We would have ordered the food even though we had no earthly reason to think he could supply it.

If he *did* supply it, that would be a miracle.

The faith of all those people walking into the restaurant with us is part of what draws us in. We count it as evidence, even though we haven't seen the goods that we seek. Whether theologian or experimental scientist, we cannot do our job unless we trust in something that we can't prove, something that we can't sense. The order of the universe, the testimony of others, the intimate contact with indescribable beauty - that is the source of our science.

9. But doesn't science eventually find its errors while religion never admits to any?

Scientists will sometimes argue that science is self-correcting: all serious errors are eventually found. It may take them a hundred years, but they are found. That is a faith-based position. It cannot be tested, cannot be demonstrated, essentially it cannot be proven.

After all, if we haven't found the error in an idea, we think the idea is true until evidence contradicts it. But what if our methods are such that the contradictory evidence exists, but cannot be detected? In that case, we will hold that erroneous concept forever because science will, by definition, be unable to detect the error.

To use an example, let us assume that science and technology were able to create a rope net with holes two inches wide. Assume likewise that despite our best formal models, our best experimental work, our best applications of this knowledge, we were unable even to conceive of a net with a smaller mesh, much less build such an object. So, we build our finest two-inch mesh net and cast it into the sea to haul up fish. We notice that every fish we catch is larger than two inches. We conclude that there are no fish smaller than two inches.

But I cannot prove the negative. I cannot use my inability to catch one-inch fish as an argument that one-inch fish must exist. Or, to use a more popular example, if I developed the idea of a Flying Spaghetti Monster, it would be absurd to argue that, since I have no net capable of catching him, such a monster must exist.

We cannot prove a negative, but we must equally recognize that there may well be more to the universe than is dreamt of in our experimental science, Horatio. Experimental science has conceptual limitations. Indeed, the whole reason we do experimental science is because we know there are things we don't know and we're trying to capture those things in our nets.

Science captures data. Philosophy interprets data. Data interpretation is not experimental science. For example, one of the classic problems in math is to draw three points on a grid and ask, "How are these points connected?" The simple answer is: with a straight line. But do we know that the correct answer is actually a straight line? Perhaps the actual line is very curvy indeed, a sine wave, for example, that loops down through the first point, up to connect the second, then loops even farther up before coming back down to connect the third. Or, worse, perhaps the actual line actually loops up and down twice between each point. Or perhaps three or four times between each – we don't know. As any decent mathematician will readily point out, there are, in fact, an infinite number of lines that can connect any point set we can come up with.

Every experiment creates a point of data. Given any set of experimental points, we can construct an infinite number of theories to account for all of those points. We tend to choose the theory that seems simplest to us, but that doesn't mean it is really the simplest or really the most accurate. It's just the answer we currently use.

Experimental science can't get beyond this problem of how best to interpret the points because it is concerned with creating the points. Only a formal philosophical system can really connect the points. Experimental science is not a formal system. It's just a data capturing tool.

10. Do math and other formal sciences fail?

Yes, formal logical systems do fail as well. This was demonstrated in a most startling way by Kurt Gödel in 1931. This brilliant Austrian mathematician showed that any logical system complex enough to produce a statement like "1+1=2" had to contain ideas which could not be demonstrated to be true or false within the same system.

Now, the idea that "1+1=2" may sound trivially easy to demonstrate, but it has more to it than meets the eye. In 1913, two eminent mathematicians, Bertrand Russell and Alfred North Whitehead, wrote a three volume work intended to demonstrate the basic ideas and basic logic underlying arithmetic. They began with the most simple axioms and showed how the interaction of these axioms led, bit by bit, to various arithmetical ideas. On page 86 of the second volume, they were able to finally conclude "1+1=2." This statement had a footnote that read, "The above proposition is occasionally useful." That gives you a feel for how much careful thought has to go into creating a logically consistent formal system.

> Einstein's religion [was] more abstract, like Spinoza and Indian philosophy. Spinoza's god is less than a person; mine is more than a person; because God can play the role of a person.
>
> ### Kurt Gödel
>
> *Mathematician and Lutheran.*
> *His Incompleteness Theorems revolutionized every logic-based discipline known to man. He used modal logic to demonstrate that Anselm's proof for God's existence is valid.*

Less than twenty years after Russell and Whitehead published this exhaustive (and exhausting) work, Gödel demonstrated that complex logical systems, such as simple arithmetic, supply answers which must, in some sense, be accepted on faith, since there is no way to use the system that produced the answer to also prove the answer absolutely has to be true.

In short, no complex logical system is self-verifying. Sometimes, one kind of logical system can be used to verify another kind of logical system, but that proof assumes the proving system is true as well. So, both experimental science and formal science are ultimately faith-based systems. They work, in part, because we have reason to trust, to believe.

11. If religion is so wonderful, why have there been so many religious wars? Why do we have to have a separation of church and state to keep the peace?

First, does taking religion out of the public sphere *really* keep the peace? Are secular countries *really* more peaceful than religious countries? War, as has often been noted, is hell. But what if you got to take weekends off?

The Catholic Church long insisted on exactly this kind of holiday. Between 989 AD and roughly 1250, war could only be waged between sunrise Monday and sunset Wednesday. No one could do violence nor confiscate anyone's goods during the four weeks of Advent and the octave of the Epiphany, or during Lent and the octave of Easter, or during the two weeks prior to or the week following upon Pentecost. Anyone who violated these days of peace was exiled for thirty years and excommunicate.

Such were the rules imposed by the Peace of God and its close relative, the Truce of God. Now, keep in mind, peasants could not be drafted for armed service when they were needed in the field to sow and reap the harvest. In a subsistence level economy, part of the spring and fall were already off-limits to war if only because inattention to the fields meant starvation for the whole land within a year.

So, when the Truce of God and the Peace of God are taken together with the natural disinclination to wage war during sowing and harvest seasons, we see a remarkable result: for medieval Christians, only 80 days of each year were left for fighting. Can anyone imagine Alexandar the Great succeeding under such terms? Even the famous Jewish reverence for the Sabbath did not restrict warfare to this degree.

No other religious tradition, Protestant or Orthodox, Muslim or Hindu, Buddhist or civilized secular scientist, ever attempted such a thing. In short, it isn't clear that religious countries are less peaceful. Now, are religious wars especially heinous? Let's check the body counts:

Wars of science (waged to support specific biological, political, economic, etc., principles)

- World War II – 72 million, including 25 million military deaths.
- World War I – 40 million total, including 9.7 million military deaths.
- Franco-Prussian war – 750,000, including 250,000 military deaths.
- Napoleanic wars – Estimated 6 million dead.
- French Revolution – 40,000 in the Revolution; probably 500,000 killed in the Vendee region.
- American Revolution – estimated between 50,000 and 100,000 dead total.
- Seven years' war – estimated 1.3 million dead, including 700,000 military deaths

Deadliest Religious Wars

- War on Terror – a minimum of 70,000 military deaths to date.
- Thirty Years' War – 7.5 million dead.
- Hungary's Peasants' War – 70,000 dead.
- Germany's Peasants' War – 100,000 dead.
- Crusades – Total deaths from 1096 to 1270 are estimated at about 1.5 million.

When we get past our preconceptions and actually study the numbers (as good scientists should), we see wars of Christian religion actually tend to be relatively low-violence affairs. Even our current War on Terror - which is not a war of Christian religion - sees only a handful of people killed every few weeks. Traffic fatalities tend to be higher. Even the deaths resulting from the destruction of the World Trade Center barely matched the number of abortions committed on September 11.

Communism and socialism are the purest of economic theory, yet the rulers of communist and socialist nations have imprisoned, maimed and slaughtered more people than the last two world wars combined. According to the UN Genocide convention, Stalin alone killed 62 million people, slaughtering more Ukrainians then his nearest mustachioed rival.

Likewise, Mao Tse-tung produced 3 million body bags in his first five years in power. But he didn't stop there. The millions who died in subsequent years, especially during Mao's Great Leap Forward, the economic policy that drove huge sections of the countryside to starvation and cannibalism, will probably never be known. Pol Pot only killed a million, but he did as well on a per capita basis as either of his more well-known rivals. And, of course, we can't leave out the most well-known economist of all, the national socialist Adolf Hitler. Though he started well, he was relatively ineffective, killing only twelve million in death camps during the war and several hundred thousand in the years leading to it.

> Every Communist must grasp the truth, political power grows out of the barrel of a gun... The seizure of power by armed force, the settlement of the issue by war, is the central task and the highest form of revolution. This Marxist-Leninist principle of revolution holds good universally, for China and for all other countries.
>
> **Mao Tse Tung**
>
> *Political theorist and atheist.*
> *He successfully fought Chiang Kai Shek for control of the Chinese mainland. He is responsible for the deaths of untold millions.*

It is, of course, unfair to limit ourselves to the 20th century. Essentially every European war since the mid-1700's has been driven in whole or part by the science of economics. The French Revolution guillotined tens of thousands, millions more died during the Napoleonic wars that swept the whole of Europe. The American Revolution was about corporate control; whether or not the East India Company could run the colonies. The Civil War tested the economic viability of slavery. England killed tens of thousands of Chinese in the Opium Wars, a kind of reverse drug war in which the British solved their trade imbalance with China by forcing China to become a nation of opium addicts — the British controlled the opium trade, you see.

So, if we keep religion out of politics to prevent war, should we not also erect a wall of separation between economics and the public sphere? After all, economics and politics are a volatile mix, as two centuries of decaying corpses illustrate. But wait. There's more.

What about biology? Ignore the dissected frogs and newts, concentrate solely on severed human heads. Darwin published *On the Origin of Species* in 1859 and *The Descent of Man* in 1871, just a year after the outbreak of the Franco-Prussian War. Most people don't realize just how strongly Darwin's work influenced European intelligentsia. In the dozen years between 1859 and 1871, the ruling class of Europe embraced the "survival of the fittest." They spoke of the "French race," the "British race," and the "German race." The Franco-Prussian War was seen on all sides as an expression of scientific principles, a necessary Darwinian struggle. Of course, that war led directly to World War I which itself led directly to World War II. Indeed, many historians refer to this three-fold slaughterhouse as the modern Thirty Years' War, our hymn of praise to the science of biology.

John Dwyer, in his remarkable work *War Without Mercy*, demonstrates that World War II was just as much about race as the Franco-Prussian War. The Japanese saw the Chinese as inferior; Americans saw the Japanese as inferior; and Germans saw Jews and Slavs as inferiors. The rationale that allowed Auschwitz also built American internment camps for West-Coast Japanese citizens and prompted FDR to suggest that the entire German people should be sterilized. This was the thinking that allowed American scientists to commit medical atrocities on Americans: the poor, the black, the mentally retarded.

So, obviously, biology and politics are a volatile mix. Federal grants for biological research, like tuition vouchers, violate this much-needed wall of separation between science and state. The wars of biology have built a much higher pile of corpses than all the wars of religion combined, while the internal and external wars of economics dwarf them both. If religion is banned from the public sphere, can anyone really argue the others can safely stay?

Why did Thomas Jefferson think a wall of separation between church and state was needed? Because the American Revolution happened two hundred years too soon. Thomas Jefferson was a well-read man, but he was completely ignorant of evolutionary theory. How could he be otherwise? It wouldn't be invented for another fifty years. He knew Adam Smith's *Wealth of Nations*, but knew not a thing of what Marx would write twenty years after Jefferson died. Besides, Jefferson was the man who re-wrote the Gospels by taking out all references to miracles. He didn't believe in them, you see. He thought Jesus was a nice moral teacher, but not God. C.S. Lewis was not yet a gleam in his father's eye, so Jefferson was unaware of the Liar, Lunatic, Lord argument, and he wasn't smart enough to figure it out for himself.

Because economics was just barely becoming a science, Jefferson could not fully appreciate to what extent science contributed to the widespread carnage wreaked even in his own day. Similarly, biology was not yet on a scientific footing, and its potential to promote slaughter was not appreciated by America's founders.

Jefferson and his associates wanted a wall between religion and politics because, as far as they knew, religion was the only known cause of major war. Take away religion, they reasoned, and you take away the major reason for conflict. They didn't realize that religion was the language of war for the simple reason that it was the only mature system of thought Europeans possessed, the only scientific language articulate enough to speak. No other system of thought had been sufficiently developed to manage human affairs. Thus, when someone wanted to justify a war, a pillage, any sort of carnage at all, they were forced to use the only words available — religious words.

But here's the interesting thing. After Jefferson's death, as other systems to manage facts developed, people could choose how to justify war: they could continue with Christian religion or they could adopt scientific rhetoric. As even a simple study of armed conflict since the Enlightenment shows, the movers and shakers of the Western world chose decisively in favor of science. They have consistently used the language of science to justify their aggression, not the language of Christian faith. The reason is simple: people like power. Christian faith does not lend itself to war. As soon as other languages were strong enough to bear the burden, "Christian rationales" dropped away.

This rejection of religious rhetoric is peculiar to the Christian cultures which developed scientific languages to replace it. For whatever reason, religious phrasing and phrases are still the preferred language for war in many parts of the world. It is worthwhile to note, however, that these religious rationales for violence almost never employ words of Christian faith. The French Revolutionaries invoked "the Goddess of Reason," jihadists invoke Islam,

Indians invoke Hinduism, and Westerners use the priests of various sciences, but progressives who seek a firmer road towards mass murder have abandoned Christianity. For them, Jesus Christ is not the door. He is a wall that blocks their road.

For the scientist who has lived by his faith in the power of reason, the story ends like a bad dream. He has scaled the mountain of ignorance; he is about to conquer the highest peak; as he pulls himself over the final rock, he is greeted by a band of theologians who have been sitting there for centuries.

Robert Jastrow

Astronomer, physicist, cosmologist and agnostic.
He was the founder of NASA's Goddard Institute and is now
director of the Mount Wilson Institute and its observatory.

CPSIA information can be obtained at www.ICGtesting.com
Printed in the USA
LVOW071037170112

264199LV00001B/6/P